The Economics of International Environmental Cooperation

Polish Studies in Economics

Edited by Ryszard Kokoszczyński

Volume 3

PL ACADEMIC RESEARCH

Tomasz Żylicz

The Economics of International Environmental Cooperation

PL ACADEMIC RESEARCH

Bibliographic Information published by the Deutsche Nationalbibliothek
The Deutsche Nationalbibliothek lists this publication in the Deutsche Nationalbibliografie; detailed bibliographic data is available in the internet at http://dnb.d-nb.de.

Library of Congress Cataloging-in-Publication Data
Zylicz, Tomasz, author.
The economics of international environmental cooperation / Tomasz Zylicz. -- First edition.
pages cm. -- (Polish studies in economics, ISSN 2191-8848 ; volume 3)
ISBN 978-3-631-65233-6
1. Environmental protection--Economic aspects. 2. Environmental protection--International cooperation. I. Title. II. Series: Polish studies in economics ; v. 3.
HC79.E5Z95 2015
333.72--dc23

2015001784

This publication was financially supported by the University of Warsaw.

ISSN 2191-8848
ISBN 978-3-631-65233-6 (Print)
E-ISBN 978-3-653-04849-0 (E-Book)
DOI 10.3726/ 978-3-653-04849-0

PL Academic Research is an Imprint of Peter Lang GmbH.

Peter Lang – Frankfurt am Main · Bern · Bruxelles · New York · Oxford · Warszawa · Wien

This publication has been peer reviewed.

www.peterlang.com

Table of Contents

Introduction 1

1. The Victim Pays Principle? 9

1.1 The Rhine Treaty 9
1.2 Issue linkage 12
1.3 Specific watershed cases 14
1.3.1 The Danube 15
1.3.2 The Nile 16
1.3.3 The Mekong 17
1.3.4 The Plata 18
1.4 Managing international rivers 19

2. Regional cooperation 23

2.1 Acid rain in Europe 23
2.1.1 The Thirty Percent Club 25
2.1.2 Protocols to the Geneva Convention 32
2.1.3 Non-European acid rain 35
2.2 Baltic Sea eutrophication 36
2.2.1 The Baltic Sea predicament 36
2.2.2 Gdansk and Helsinki Conventions 39
2.2.3 Prospects for a regional policy 45
2.2.4 Implementation of hypothetical transfer mechanism 50

3. Protecting the Global Commons 53

3.1 Ozone layer 53
3.1.1 Ozone depletion story 53
3.1.2 DuPont *versus* ICI 56
3.1.3 The Montreal Protocol 58
3.2 Climate 64
3.2.1 The greenhouse effect as a public good 65
3.2.2 The Intergovernmental Panel on Climate Change 70
3.2.3 The Berlin Mandate 73
3.2.4 Carbon leakage 79

3.2.5 Will co-benefits help? 83
3.3 Biodiversity preservation 86
3.3.1 Biodiversity as a public good 87
3.3.2 The 1992 convention on biological diversity 88

4. Cross-cutting Issues 95

4.1 International development assistance 95
4.2 The institutional framework for environmental protection 99
4.2.1 Debt-for-Nature swaps 99
4.2.2 The Polish EcoFund experience 106
4.2.3 Applied game theory 116
4.2.4 Lessons learnt from the Polish debt-for-environment swap 121
4.3 Environmental aspects of European non-environmental policies 123
4.4 Trade and environment 130
4.4.1 Academic analyses 131
4.4.2 Policy conclusions 134

Summary and Conclusions 137

References 141

Introduction

"Nature does not respect political boundaries". This is no doubt a true statement, but its real meaning is vague. To be sure, nature cannot be appropriated by a single nation. But what are the practical consequences of this fact?

This book on international environmental cooperation confronts two points of view: natural and economic. According to the natural viewpoint, all borders – so carefully protected by us – are artificial and damage the cause of environmental protection rather than further it. On the other hand, according to the economic viewpoint, borders are necessary; they are inevitable elements of our political systems. They allow societies to enjoy stability and maintain justice. Within borders, we try to make sure that foreign citizens do not draw from our pension systems, and we pressure governments to financially and politically support companies that export products.

The co-existence of these two orders is inevitable. Yet it is worthwhile to analyse the ways in which sovereign entities – such as nation states – join forces when solving problems that cannot be solved locally, i.e. problems that call for cross-border cooperation.

Many governments and international organizations extol the *Subsidiarity Principle*, which states that decisions should be taken at the lowest possible level that is adequate given the nature of a so-called *externality* problem. There would be little sense in asking an international organization to protect the landscape in any single given location; local authorities are capable of coping with such problems. By contrast, effective protection of the Baltic Sea requires the cooperation of the Baltic states; effective protection of the air in Europe requires the cooperation of the entire continent; and protecting the ozone layer cannot be effective unless all countries cooperate.

International environmental cooperation is therefore inevitable for solving some problems. It is necessary when local solutions are either impossible or inefficient from a global perspective. This impossibility is linked to what economic theory refers to as *public goods*. If a piece of the natural environment cannot be appropriated by a single region, then a sort of a "joint responsibility" applies. One can do something about this piece, but the outcome depends on what others do. The lack of effectiveness of climate protection measures in the absence of a global agreement to limit greenhouse gas emissions serves as the best illustration of the point in question. Likewise, solving the "acid rain" problem by erecting higher stacks is an example of a relatively inexpensive method of lowering local sulphur and nitrogen depositions at the expense of neighbouring countries. Economists see this as a textbook externality case: overall efficiency is compromised by transferring damage from one place to another.

Economic efficiency – synonymous with maximizing a positive difference between benefits and costs – remains an important focus of economic analyses. Seen from such a point of view, international environmental cooperation is not an entirely new domain of economic inquiry. Nevertheless, what requires special attention is the asymmetry between benefits and costs perceived by sovereign states as economic agents. This asymmetry can be spotted at the local level, too. In a typical polluter-pollutee system, the former gets rid of a problem by transferring it to the latter. In a typical public good system – e.g. protecting a forest – a protecting agent finances a project from which other agents benefit; the latter may nevertheless pretend that they are not interested in protection, especially when they know that this protection is financed by somebody else, which means that they can benefit for free. In such cases, a solution can be implemented by the government that enforces what is justified by economic efficiency and/or fairness.

Environmental economics is about striking a balance between protection costs and benefits. Under standard assumptions, this balance is reached when the marginal costs of abatement are equal to its marginal benefits. The benefits may be enjoyed by agents other than those who incur the costs. But the role of the government is to make sure that the difference between total benefits (whomever they accrue to) and the total costs (whoever pays them) is maximized, as envisaged on a figure that can be found in any textbook on environmental economics. Costs and benefits (p) are measured along the vertical axis, while the level of environmental protection (e) is measured along the horizontal one. The low level of environmental protection e^* is what can be expected if polluters are free to do whatever is most convenient for them, i.e. to avoid any costly activities. In contrast, e^0 is what maximizes the difference between total abatement benefits (TAB) and the total abatement costs (TAC). Under standard assumptions, this is where marginal abatement benefits (MAB) are equal to marginal abatement costs (MAC).

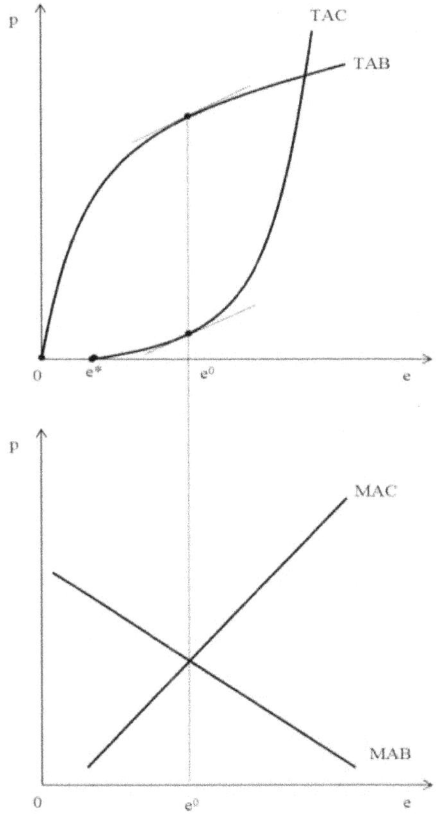

Yet in situations when environmental protection requires the cooperation of sovereign countries, there is no government that can mandate anything. And this is the core of international environmental cooperation. Theoretical analyses demonstrate that a solution can be arrived at by reducing external costs or undertaking common endeavours for the common good. But how can this be achieved? How to encourage a sovereign country to make an effort that it would not undertake spontaneously?

Some steps can be enforced by a military action. Over the last several thousand years, there have been wars aimed at forcing somebody to do something that they would have not undertaken voluntarily. Analyses exist that interpret wars fought during the last decades as wars for cheap energy. Some analysts predict that wars in the 21st century will be fought over access to fresh water. Thus, it is possible to look for environmental motives of some wars, though the actual implementation of environmental protection usually does not require military force.

Choosing a socially justified environmental protection level requires comparing broadly understood costs and benefits. The analysis may demonstrate that what is justified for a group of countries may not be justified for a single country. The very definition of optimality implies that the surplus of net benefits of winners is higher than the deficit of net benefits of losers. Consequently, the winners can afford to compensate the losers and still be better off. This justifies international environmental cooperation supported by compensating the losers by the winners in order to jointly undertake something that is beneficial for all, which is a variant of a general principle of environmental policy in a single country where the government enforces solutions justified by a surplus of total benefits over total costs. This does not necessarily imply that winners pay compensations to losers directly. The losers may tolerate their fate because they appreciate the benefits of operating in a country where somebody else wins.

Indeed, some examples of international environmental cooperation are linked to compensations. These do not have to be direct or transparent. There are cases where a sovereign state accepts an environmental commitment which implies a net loss. In order to understand this paradox, one needs to analyse the broader context of a particular agreement and presumably discover that, at the same time, the state negotiated another agreement where it was a winner; thus, the two decisions should be studied together.

Sometimes we encounter *self-enforcing agreements*. Parties to such agreements find it more beneficial to stay rather than to move out of a respective coalition, which is how analysts explain the success of the Montreal Protocol protecting the ozone layer, or the participation of some countries in the Polish Debt-for-Environment Swap. Losses from leaving a coalition do not have to be linked to economic (or military) sanctions; they may simply be linked to losing benefits that come with participation. How to bring together a "critical mass", i.e. to collect enough participants so that the agreement becomes self-enforcing, remains an open question. This, however, depends on particular environmental problems.

There is an abundance of excellent literature on environmental economics and policy. While Baumol and Oates (1988) serves as a classical reference for policy analyses, Hanley *et al.* (2007) can be recommended as a good overview of the most salient economic aspects.

Environmental economics is often linked to natural resource economics. Many texts and courses cover both of them. Indeed their fields overlap and so do their methods. Environmental quality is sometimes considered a natural resource, which makes one a part of the other. Despite that, many economists argue that environmental economics is about externalities, and that resource economics is about trajectories over time.

There are two broad categories of natural resources: exhaustible and renewable. The former – comprised of fossil fuels and other raw materials – includes goods whose inventories have to be depleted once people use them. The latter can be naturally reproduced over and over again if people use them reasonably. Timber is a prime example of this category. Some analysts introduce a distinct category of non-depletable resources – such as e.g. salt (NaCl) – whose supply is virtually infinite, given the demand. Economists are not concerned about such resources, since no difficult trade-offs are involved.

Of course, using natural resources may create external effects. For instance, pumping oil from a deposit that is also used by somebody else may imply an extra cost for the other agent. Likewise, catching a fish may create an external cost for other fishermen – now or in the future. To the extent that using natural resources does trigger international cooperation, it will be addressed in this book. Some of these endeavours can be analysed from the point of view of managing a raw material, but they primarily are of interest as initiatives aimed at controlling environmental quality.

Many economists downplayed the uniqueness of the raw materials market, arguing that all markets – including labour, technology etc. – are essential for people's well-being (Olson, 1963). On the other hand, some raw materials, whose supply is concentrated in few countries, can be used strategically by the suppliers (Mikdashi, 1974). Nevertheless, it is highly doubtful whether the OPEC success can be replicated in another natural resource market. So far it has not been replicated.

Natural resource scarcity has been disturbing economists at least since Malthus (1798), who anticipated a rapid decline in growth caused by an absolute barrier. Ricardo (1817) observed that, before the Malthusian barrier is encountered, a decline in resource quality – resulting in their rising prices – will impede growth gradually. Yet Mill (1848) concluded that both the Malthusian (absolute) and Ricardian (relative) barriers to growth are successfully overcome by technological progress. This optimistic assumption allowed economists to disregard natural resources for several generations.

In the middle of the 20th century, natural resource scarcity started to concern politicians and economists once again. The US government appointed the so-called Paley Commission to study "the broader and longer range aspects of the nation's materials problem as distinct from the immediate defense needs." Its report outlined a number of problems, but the commission did not find a sufficient evidence for major concern. Two American economists – Barnett and Morse – undertook a gigantic effort to verify the Malthus-Ricardo-Mill hypotheses against available empirical material. Overall, they found no evidence of growing resource scarcity (Barnett, Morse, 1963) and therefore no need for a specific policy in this area.

The issue has been addressed many times. At the beginning of the 21st century, Resources for the Future (RFF) – an American think tank established as a result of the Paley Commission – published its *Scarcity and Growth Revisited* report (Simpson *et al.*, 2005). The report points at environmental quality as the most cumbersome issue resulting from natural resource extraction. Thus, not scarcity as such, but rather protection of the environment should be a matter of concern for policy makers, and a focus of research for economists.

An interesting case is provided by pressures on the Chinese government to open its rare earth metal extraction to international trade. China has around a third of the world's deposits, and accounts for virtually all of their international trade presently. As a monopolist, China is accused for strategically restricting the supply of rare metals in order to elevate their price. Chinese authorities deny these accusations by pointing out that export quotas are much higher than actual trade. However, it is estimated that domestic prices are 36% below reported FOB prices (Els, 2014), which has been caused by extremely high export taxes. On the other hand, Chinese authorities argue that they aim at lower production not because of monopolistic strategies, but because of extreme environmental disruption caused by extraction. Nonetheless, as the disruption is local in scope, foreign governments are concerned about supply rather than extraction technology.

In its 2014 ruling, the WTO (2014) indicated that Chinese regulations do discriminate exports *vis à vis* domestic production. It is yet to be determined whether China will exert its political power to maintain constraints placed on its rare earth exports or will make its environmental policy export-neutral. In any event, this shows that international environmental cooperation can be linked to natural resource policy questions.

In 2011 the European Commission published its *Roadmap to a Resource Efficient Europe* (EC, 2011), which calls for lowering raw material requirements in production. While such an improvement seems inevitable, there were attempts to elaborate detailed mandatory targets and price controls. The latter would have been unfortunate. Motivated by high price volatility, price controls would paradoxically serve against resource efficiency. Controls are supposed to lower the price if the market boosts it "too much" and to make it higher otherwise. This is to be done by adjusting taxes. Proponents of the mechanism claim that tax authorities have greater knowledge of scarcity than market agents. This, however, is disputable, and price controls would most likely serve as a mechanism shielding economic agents against the market, sending signals about resource scarcity.

Even though natural resources are a matter of concern for international bodies, environmental protection has been a much more popular motive of cooperation. Thus, we will focus on environmental rather than resource policies. Externalities serve as a reference for economic analyses of environmental cooperation problems (Mäler, 1990). In the case of global externality, one can differentiate between situations in which many sources accompany few victims, many victims accompany few sources, or – the most complicated one – the environment is a common property.

This book consists of four chapters. The first chapter reports on cases where the *Victim Pays Principle* (rather than the celebrated *Polluter Pays Principle*) seems to be adopted. Possible indirect compensations are discussed whenever the distribution of benefits and costs seems to be evidently asymmetric. In the second chapter we look at regional cooperation tied to solving problems such as Acid Rain in Europe or Baltic Sea eutrophication. Addressing global problems – mainly the ozone layer and climate, but also biodiversity – is the topic of chapter three. In particular, we contrast the success of the Vienna Convention with the failure of the Rio Convention (while the former focuses on the ozone layer, the latter focuses on climate). Chapter four looks at some cross-cutting issues, including development assistance and trade.

1. The Victim Pays Principle?

This is not a joke. Environmental analysts have long been familiar with the *Polluter Pays Principle.* Virtually all governments declare this principle as the cornerstone of their environmental policies. Defined by the OECD in 1972, it says that the polluter is responsible for whatever harm the pollution imposes (Barde, 1992). Quite often it turns out that it is difficult to identify a polluter, and the principle has to be compromised, but at least in its narrower form – stating that the polluter is responsible for meeting whatever environmental requirements are imposed by authorities (redefined by the OECD in 1974) – it has been almost universally accepted. Even though the *Polluter Pays Principle* has been broadly considered the cornerstone of environmental policy, the OECD (1981) admitted that it does not necessarily apply to transboundary problems.

The Coase (1960) theorem looks at the *Polluter Pays Principle* as a non-obligatory condition to achieve economic efficiency in environmental policy. Indeed, there is a sort of symmetry between those who generate externalities and those who are affected. It can well be conceived that the victim pays the polluter a compensation for scaling the pollution down to a socially justified level. Even though media and public opinion do not accept such a result, this is sometimes how environmental policies are carried out. Government subsidies to sewage treatment plants are a prime example of this sort.

As a rule, however, solving an environmental problem implies that the polluter is implicated in one way or another. The solution rests on the assumption that there is an agent authorised to enforce what turns out to be justified on economic grounds. In relations between sovereign countries there is no such an agent. The United Nations is a group of sovereign states and key decisions are supposed to be agreed upon by everybody. Even in the European Union, where the Commission (playing the role of a supra-national government) is given a fair degree of power, many decisions are to be taken unanimously. Thus, international environmental cooperation cannot replicate modalities applied in national environmental policies.

1.1 The Rhine Treaty

In 1986 a catastrophic fire in the Basel-based chemical giant Sandoz reversed the process of bringing life back to the Rhine river. Later analyses suggest that the mess which accompanied the rescue action gave opportunities for many other chemical companies to discharge their troublesome inventories as well. Hence,

Sandoz is not the only party guilty of the disaster. What is important is that the Basel fire brought the long process of the Rhine clean-up to a stop.

The Rhine is one of the best known European rivers. According to its length, its rank is 11th, but it is almost as international as the Danube. Before it enters the Netherlands, it flows through Switzerland, Germany and France. Its drainage basin overlaps with Austria and Luxemburg (as well as – marginally – with Belgium and Italy). Therefore many countries affect the quality of its water. International protection of the river was initiated as early as in 1815 (at the Vienna Congress), but more substantial cooperation was triggered by the Bern Convention in 1963, and by two 1976 agreements dealing with Chemical and Salt (Chloride) pollution (Frijters and Leentvaar, 2003).

Originally, it seemed impossible that the Rhine – with its gigantic flow (on average more than 2000 m^3/s in its lower part) – could be contaminated. Nevertheless, by the end of the 19th century, it succumbed as a result of industrialisation of its drainage basin. Concentration of dissolved oxygen kept falling until the 1970s. Destroyed river habitats were later restored. Cadmium contamination was eliminated as late as the 1980s. Also the efficiency of nitrogen and phosphorus abatement in sewage treatment plants was improved significantly as late as the 1980s.

Concentration of chlorides, discharged mainly from French mines, was rising until the 1970s. In principle it was regulated by the 1976 agreement, but progress was slow. A more effective control resulted from the Salt Treaty signed in 1991, which protected the Netherlands from water that not only was not potable, but was also unsuitable for agricultural uses. The Dutch were the party most interested in solving the problem, though they were not responsible for the river contamination. According to the *Polluter Pays Principle*, the Dutch could have expected that the protection would be paid by other parties. Yet this has not happened.

The costs of reducing chlorides in the Rhine river were to be financed by the Saline Funds, whose budget was several tens of million euros per year. Contributions to the Fund came from the polluting countries: Switzerland (6%), France (30%), and Germany (30%), and from the Netherlands as well (34%). Hence, the victim was supposed to pay more than any of the polluters individually. Budgets of other Rhine conventions – definitely lower than the Saline Fund – take the Netherlands as their main contributor.

Economists who see the *Polluter Pays Principle* as the cornerstone of any reasonable environmental policy can challenge such solutions. Perhaps many Dutch citizens look at the Rhine with disgust and do not see a reason to pay for what should have been paid by somebody else. Yet they had to choose whether to wait until German, French and Swiss industries were forced by local governments to abate adequately, or to take the lead and pay for the clean-up in order to avoid even higher losses. What other instrument did the Dutch have to force the Germans and the French to cooperate?

The Rhine example demonstrates that international environmental cooperation is subject to different rules than environmental protection in a single country. In a national case, there is an authority to establish a justified level of protection and – more importantly – to implement enforcement actions. In an international case, there is no enforcing agent even if costs and benefits are assessed adequately. What if the Dutch were able to estimate their losses and determined that these are higher than abatement costs upstream? Would the Germans and the French conclude about the necessity to abate?

No, they would not have to. But if the victim offered to co-finance the necessary abatement activities, the balance could change. The cost to be born by the polluter goes down. This alone can change its behaviour. Perhaps the abatement cost to be born by the polluter becomes so low that local benefits alone justify the clean-up. Chlorides in the middle Rhine hurt the Germans and the French as well. The Dutch are not the only victims.

Rhine protection is not an exception. Victims contribute financially to many other international environmental programmes. The best known example of this sort is the Chernobyl disaster. It resulted in high expenditures on damage remediation and on making the crippled reactor safe. The majority of these expenditures were financed by victims rather than the Soviet Union. International programmes to save biodiversity in tropical countries are also financed by regions where people seem to be concerned with biodiversity loss rather than by those countries where species become extinct. Conversely, if a country responsible for damages participates in their healing, it insists that this has nothing to do with accepting responsibility. For instance, this phenomenon was clearly visible in 1989–1990, when the Czechoslovakian government agreed to pay for some flood damages (in Poland) caused by negligence in the upper Odra river basin (in the Czech Republic), but maintained that the payments cannot be called "compensation".

From the economic theory point of view, a victim's contribution towards environmental protection costs is not precluded. The Coase theorem explains why it does not have to violate economic efficiency. Indeed it can be traced in some national environmental programmes. In international programmes, it has become a typical way of solving externality problems.

1.2 Issue linkage

The downstream-upstream relationship is fairly simple, but many environmental problems reveal a more complicated nature. Often externalities happen to be mutual ones; victims have opportunities to bite back. Anticipating revenge, polluters may undertake some abatement actions without being compensated by victims. International environmental cooperation provides examples of that sort, too. In other words, a sovereign country accepts a commitment which seems to imply higher costs than benefits from its particular point of view.

A prime example of such a strange behaviour is cooperation between the USA and Mexico over the Colorado river (Hundley, 2009). According to knowledge acquired in primary school, the mouth of the Colorado is in the California Gulf of the Pacific Ocean. This knowledge reflects the reality of the 19th century. In the meantime, however, water abstractions made the Colorado river disappear before it reaches the US-Mexican border. It simply sinks in the sand of a desert. Tourists who are overwhelmed by the beauty of the Colorado flowing through the Grand Canyon do not have to know that downstream its water will be used by millions of people living in several cities in Nevada and California, and – even more importantly – it will be used for agricultural irrigation. As a result, the river that once was a symbol of the power and dynamism of the American West, ends up as an unattractive, shallow pond.

Consuming water for their municipal and agricultural needs, Americans serve their own interests. The fact that Mexico suffers its deficit is a typical externality. The US-Mexican border was delineated in 1848. In 1884 the International Boundary Water Commission was established, which was supposed to handle externalities arising along all three cross-border rivers (not only the Colorado, but also the Tijuana and Rio Grande). The Commission was not particularly energetic, which – given the huge asymmetry of interests and economic potential – is not surprising at all.

What may be surprising is that, in 1944, the US and Mexico signed a treaty granting the latter the right to abstract 1850 m^3 annually from the Colorado river. This is not a huge amount of water, as it corresponds to 60 m^3/s, i.e. merely 5% of the flow in the lower Vistula. Given the fact that the Colorado is twice as long, and has a large water basin, the allocation granted by the treaty was not a very generous one. And even this amount was highly theoretical, as Mexican authorities complained about how the US government complied with its obligations. Though the guaranteed flow was perhaps achieved, the liquid crossing the border with Mexico had a concentration of chlorides which precluded any economic use. After three decades, in 1974, a new treaty was signed where the USA committed not only to a flow, but also to a decent quality of water.

Additional negotiations were carried out until 2012. The most recent agreement gives Mexico much better rights to the Colorado water than before. A particularly important right is linked to the Mead water reservoir located in Nevada. Mexico is entitled to 307 million m^3 from this lake (assuming that its water level is between 322 and 343 m above sea level). Moreover, Mexico does not have to use the water for its own purposes; it can be sold to the Americans. Prospective buyers include not only municipalities, but also farmers who own extensive irrigation systems. Additionally, US authorities will assist the Mexicans to restore the non-existing mouth of the Colorado river.

Apart from the generosity of the American government (which does not seem to be spectacular here), the mere fact of sharing resources that do not have to be shared at all is remarkable. It is rare in international relations to see a party giving away rights that could have been retained. In this case the US government spent some money in order not to expose the Mexicans to externalities. Why has this happened?

Some analysts consider sharing the Colorado water a gesture that indicates a new understanding of ecology. Yet economists look for more "down-to-the-earth" interpretations. Apparently the American government wants to favour the Mexicans. Political scientists try to discover parallel negotiations in order to calculate that the benefit and cost balance is neutral: they try to identify areas where Americans are winners rather than losers. Perhaps establishing military bases is such an area. Perhaps Americans would like to buy good relations with neighbours with whom they have numerous conflicts. By the way, sharing the Colorado water has not contributed to warming the Mexico-USA relationship. Apart from the fact that millions of Mexicans dream to cross their Northern border, anti-American feelings have been quite widespread.

Occasionally an example of American-Canadian relationships is referred to in order to demonstrate that direct net benefits are claimed by the USA (Stanley, 2012). In the second half of the 20th century, states in the Northwest – especially Oregon – suffered losses caused by floods in the Columbia river basin. The Columbia flows through the Canadian Rocky Mountains, but eventually crosses the border and reaches the Pacific Ocean in the USA. In 1961 a treaty was signed to control floods. In 1964 a protocol was added to list Canadian obligations to build water reservoirs whose beneficiaries are located mainly in the American states of Montana, Washington, Idaho and Oregon.

According to the protocol, the Canadian retention reservoirs are to store as much as 19 km^3 of water.

In the Columbia River Treaty, US negotiators were aware of the net surpluses likely to benefit Canada rather than the USA, but they agreed to this treaty in light of other political circumstances (Mäler, 1990). Likewise in the Colorado Treaty. The Americans were willing to undertake a costly desalinization project in Yuma, Arizona, for three purposes. First, the project was to be financed from the federal budget, rather than the state budget. Second, negotiations linked the Colorado river (where Mexico was the downstream party) to the Rio Grande river (where the USA was the downstream party). Third, the Americans were interested in nurturing friendly relationships with the Mexicans in general (Mäler, 1990).

Some political scientists consider the Columbia treaty an example of providing an environmental benefit in exchange for a favour in some other area. This interpretation does not seem to be correct, given that the USA paid for the retention capacity. There are complicated accounting methods to clear the cost of hydroelectricity production in the USA with the cost of water retention in Canada. But the most important factor which precludes seeing the Columbia treaty as an issue linkage is the fact that the Americans paid over 64 million dollars for the water reservoir construction. One can question whether this sum has been adequate, but the treaty resembles more a market transaction than the Mexican-American Colorado agreement.

There exist cases where environmental benefits are provided in exchange for benefits in some other area. Nevertheless a more common situation is when negotiating parties look jointly for a solution to a common environmental or natural resource problem.

1.3 Specific watershed cases

In this section we briefly address how the international community has tried to manage several transboundary river resources. In some cases, efforts are aimed at coping with externalities caused by upstream parties depriving downstream parties of water (e.g. the Nile or the Mekong). Sometimes managing an international public good like navigation (e.g. the Danube) or hydropower (e.g. the Plata) is the focus of these efforts. These are fascinating cases to study ways in which policy makers try to improve economic efficiency, though most of the analytical framework has already been provided above in discussions about the Rhine, Colorado and Columbia experiences.

1.3.1 The Danube

The Danube is the most international European river. It has ten riparian states: Germany, Austria, Slovakia, Hungary, Croatia, Serbia, Romania, Bulgaria, Moldova, and Ukraine (even though the drainage basin includes also Slovenia, the Czech Republic, Bosnia and Herzegovina, and – marginally – Albania, Italy, Macedonia, Montenegro, Poland, and Switzerland). International attempts to regulate navigation along the lower and middle Danube go back to the 17th century, and involve Russia, the Habsburg empire, and Turkey. In 1856, after the Crimean War, the Commission Européenne du Danube (CED) was established. It assumed authority to control navigation in the three canals in the Danube Delta: Kilia, Sulina, and St. George. An 1865 agreement between Austria-Hungary, Britain, France, Italy, Prussia, and Turkey made the CED an almost sovereign entity by assuming that some of its decisions taken by a majority vote are binding for member states. In 1881 Russia withdrew from the agreement, resuming its sovereign control of the Kilia's left bank (at the border with Romania). Even though CED was an almost sovereign entity, it promoted British national interests by establishing a virtually monopolistic position of a British shipping company (Furness, Withy and Co.) in the Lower Danube.

World War II marked the liquidation of CED, making the Danube "a German river controlled by Germany" (Hitler's words). In 1948 the Commission was re-established by the Belgrade Convention, although it was confined to the riparian states including – at that time – the Soviet Union and Yugoslavia (with headquarters in Budapest). Now its parties are Austria, Bulgaria, Croatia, Germany, Hungary, Moldova, Slovakia, Romania, Russia (as the legal heir to the Soviet Union), Ukraine, and Serbia. As before, its aim is to facilitate navigation rather development.

Environmental considerations emerged more recently. In 1994 the Danube River Protection Convention (DRPC) was signed in Sofia. In addition to the ten riparian states and Russia – who signed the Belgrade Convention – the DRPC was also signed by Bosnia and Herzegovina, the Czech Republic, Montenegro, Slovakia, and the European Union (whose members account for a major part of the watershed). In 1998 the International Commission for the Protection of the Danube River (ICPDR) was established within the DRPC. The ICPDR promotes awareness of specific environmental protection needs of the Danube basin, but it does not seem to be very aggressive in terms of recommending that the parties establish additional strict environmental protection standards (*Joint Statement*, 2007). It seems that environmental considerations are introduced mainly by the European Union. In 2010 the European Commission published a document, *A European Union Strategy for the Danube Region* (Communication, 2010) that emphasised that the region has rich environmental assets such as "exceptional fauna and flora, precious water resources, and outstanding landscapes" which should be "sustainably preserved and restored". The EU member states are bound by specific water protection regulations – in particular the Water Framework Directive (2000/60/EC). Beyond that, the five non-member states agreed to implement the directive throughout the whole basin within the frame of the DRPC. In addition, the ICPDR serves as a coordination platform for the basin-wide implementation of the EU Floods Directive (2007/60/EC).

1.3.2 The Nile

The Nile is considered the world's longest river (6,700 kilometres) flowing through ten countries in North Eastern Africa: Rwanda, Burundi, Democratic Republic of the Congo (DRC), Tanzania, Kenya, Uganda, Ethiopia, Eritrea, South Sudan, Sudan and Egypt. Sudan accounts for the largest part of its watershed (1.9 million km^2), but out of the four major tributaries to the Nile, three originate from Ethiopia: the Blue Nile, Sobat and Atbara. They provide 85% of the water. The Nile river is truly an international resource.

Egypt has been the most important user of the Nile water (mainly for agricultural purposes) over the last several thousand years. At the same time it has been its downstream party, susceptible to whatever externalities may be caused by upstream parties. For a long time – because of its economic and political power – Egypt managed to remain the main user of the Nile water, thus acquiring *de facto* unquestionable property rights to its flow. This, however, changed in the 19th century. Several treaties were signed in order to confirm these rights. In 1891, Italy and Britain signed a treaty committing Italy not to modify the flow of the Atbara river to lower the supply of water further down. In 1902, Britain and Ethiopia signed a treaty protecting the flow that could have been altered by some engineering works on the Blue Nile and Sobat rivers. Two treaties were signed in 1906. On behalf of Congo, the Belgian government signed a treaty with Great Britain to protect the flow to the North, thus unilaterally favouring downstream parties. In a tripartite treaty (between Britain, France, and Italy) Egypt was guaranteed a sufficient supply of water. It was contested by Ethiopia, whose water resources were in fact managed by the treaty.

Many of the subsequent Nile agreements attempted to reconcile the conflicting interests of Ethiopia and Egypt. In 1925, Great Britain and Italy exchanged notes to guarantee a sufficient water supply for Egypt. Ethiopia protested against these notes and called for intervention by the League of Nations. The conflict was soothed when British and Italian governments confirmed Ethiopian sovereignty over Lake Tana, which gives water to the Blue Nile (it is also a biodiversity treasure). Nevertheless, conflicting interests of Egypt and Ethiopia over the Nile must be included in any attempt to understand armed conflicts that broke out between e.g. Somalia and Ethiopia in the 1970s and after (Oloo, 2007).

Another conflict evolved between Egypt and Sudan. In 1929 an agreement was signed by the governments of Egypt and Sudan (which was an Anglo-Egyptian condominium at that time). The agreement was highly asymmetric, giving Egypt the right to over 90% of the annual flow, and retaining total Egyptian control over water during the so-called dry season, when water is crucial for agricultural purposes. Sudanese interests were accounted for later on. In 1959 an agreement was signed by both countries which gave 25% of the water rights to Egypt and 75% to Sudan; if other riparian states were to claim some water in the future, their allocations would proportionally subtract from what Egypt and Sudan appropriate. Both countries agreed that Egypt would build its Aswan High Dam, and Sudan would build its own hydropower/retention infrastructure. In order to reduce evaporation losses, Sudan was supposed to carry out some engineering works in the South, while the cost of these projects were to be shared by both countries. Incidentally, the work planned in the southern swamps is highly controversial from the environmental protection point of view, but in the 1950s this was not a priority issue.

1.3.3 The Mekong

The Mekong is another trans-boundary river. Its estimated length is 4,350 km, and the watershed is 795,000 km^2. The river originates in Tibet, runs through the Chinese Yunnan province, Burma (Myanmar), Laos, Thailand, Cambodia and Vietnam. The Mekong is much more difficult for navigation and irrigation than the Nile because of extreme seasonal variations in flow and the presence of numerous rapids and waterfalls. Nevertheless it has served as a major trading route from the Chinese Yunnan to the South.

Soon after Indochina was decolonized, an international Mekong Committee was established, which was strongly supported by the American government, which was afraid of possible Soviet and Chinese penetration into the region. Most of the committee's activities concentrated on developing hydropower in the region. Nevertheless, only one dam – Nam Ngum, located in Laos – can be considered an international project, since the electricity produced there is sold mainly to Thailand. In 1975 the committee adopted a declaration to prevent unilateral water appropriation without prior approval of parties affected, but as a result of the "Khmer Rouge" revolution, subsequent region-wide activities were suspended. Between 1978 and 1992 Laos, Thailand, and Vietnam cooperated within the framework of the so-called Interim Mekong Committee until an intense conflict broke out over a water diversion project in Cambodia. At that time Thailand threatened its neighbours by halting collaboration. The International community made serious attempts to resume this collaboration, which eventually led the four downstream countries – i.e. Laos, Thailand, Cambodia and Vietnam – to establish the Mekong River Commission (MRC) in 1995, which was to regulate the Mekong's resources. China and Burma (Myanmar) became "dialogue partners" of the MRC in 1996. From then on, all the six countries of the watershed have cooperated openly. It seems that the MRC abandoned the ideology of large scale hydropower development, emphasizing instead the livelihoods of local populations. In particular, it withdrew from a gigantic Pa Mong dam project (in Laos) which would have displaced more than 250,000 people. Despite that, environmental experts have expressed concerns over Chinese desires to confine MRC discussions to flood control (in order to avoid the dry-season flow issues). In addition, downstream states fear that retention reservoirs planned in the Upper Mekong Navigation Improvement programme (in China), in addition to hydropower, may be used for irrigation, which will deteriorate the dry-season flow (Sneddon, 2007).

1.3.4 The Plata

The main tributaries to the Rio de la Plata – i.e. the Parana and the Uruguay rivers – make up the second largest watershed in South America (after the Amazon) of 4.1 million km^2. The Plata itself is only 290 km long, but its width is from 2 km (where the Parana meets with the Uruguay) to 200 km, which makes some geographers consider it an estuary rather than a river. The Parana is 4,880 km long. It plays an important role in transport. Now, with the construction of several dams, its upper part is not navigable, but transport losses are considered offset by hydroelectricity production.

In 1969 all the riparian states – i.e. Bolivia, Brazil, Paraguay, Uruguay, and Argentina – signed an umbrella La Plata River Basin Treaty, though individual development and hydropower projects are worked out in separate bilateral (sometimes multilateral) agreements. The largest of these – the "Hidrovia", conceived in the 19th century – materialised in 1989 as a gigantic endeavour involving the entire watershed. Even though its aim was to make the rivers navigable, it consumed some of the earlier hydropower projects, including the famous Itaipu dam, which flooded the Guaira Falls. The latter – located right at the border of

Paraguay and Brazil – was a matter of dispute as both countries claimed the property title to the energy produced. The huge (26 GW) construction project, which provides Paraguay and Brazil with 78% and 26% of their electricity, respectively, ended the dispute when, in 1973, both countries agreed to cooperate (Da Rosa, 1983). Environmentalists are critical of some of these monumental hydropower and navigation projects, since they compromise the unique biodiversity of the region. The Parana sub-basin is considered extremely rich in fish and bird species, some of which are endemic. As a result of serious environmental doubts (as well as financial viability problems) "Hidrovia" has not progressed as planned, especially north of Asuncion in Paraguay.

1.4 Managing international rivers

Sovereign states have used international rivers to their advantage. As a result, downstream parties could suffer externalities caused either by high water abstractions or by poor water quality polluted by upstream parties. When their political advantage turned out to be insufficient, they have tried to limit damages by convening appropriate international agreements. Occasionally – as in the Rhine case – they resorted to the *Victim Pays Principle* in order to motivate polluters (or upstream users) to move towards a more efficient use of water.

Navigation is another issue. Economic theory views the problem as that of a public good. In other words, cooperation requires users to participate in financing common infrastructure. However, river navigation is not a pure public good, since potential users can be deprived of benefits by keeping them away. Activities of the Danube Commission in the 19th century provide an example of how a powerful agent (in this case, the British government) can use an unconstrained access ideology to negotiate an unfair deal (in this case with Romania) on navigation in the delta region.

The international community attempted to establish general rules on how to manage rivers. In 1992 in Helsinki, the United Nations Economic Commission for Europe (UNECE) agreed on the Convention on the Protection and Use of Transboundary Watercourses and International Lakes. The UNECE includes all the European countries and – for historical reasons – also the US and Canada. The convention was rather soft, which resulted in its universal acceptance. Its article 2.5 outlines three principles that riparian countries should apply:

- *The precautionary principle, by virtue of which action to avoid the potential transboundary impact of the release of hazardous substances shall not be postponed on the ground that scientific research has not fully proved a causal link between those substances, on the one hand, and the potential transboundary impact, on the other hand;*

- *The polluter-pays principle, by virtue of which costs of pollution prevention, control and reduction measures shall be borne by the polluter;*
- *Water resources shall be managed so that the needs of the present generation are met without compromising the ability of future generations to meet their own needs.*

In 1997 in New York, more than 100 countries signed the Convention on the Law of Non-Navigational Uses of International Watercourses. Despite fairly wide acceptance, the ratification process was rather slow. The convention came into effect as late as 2014, following its 35th ratification. It has several notable articles:

5.1. *Watercourse States shall in their respective territories utilize an international watercourse in an equitable and reasonable manner. In particular, an international watercourse shall be used and developed by watercourse States with a view to attaining optimal and sustainable utilization thereof and benefits therefrom, taking into account the interests of the watercourse States concerned, consistent with adequate protection of the watercourse.*

5.2. *Watercourse States shall participate in the use, development and protection of an international watercourse in an equitable and reasonable manner. Such participation includes both the right to utilize the watercourse and the duty to cooperate in the protection and development thereof, as provided in the present Convention*

7.1. *Watercourse States shall, in utilizing an international watercourse in their territories, take all appropriate measures to prevent the causing of significant harm to other watercourse States.*

7.2. *Where significant harm nevertheless is caused to another watercourse State, the States whose use causes such harm shall, in the absence of agreement to such use, take all appropriate measures, having due regard for the provisions of articles 5 and 6, in consultation with the affected State, to eliminate or mitigate such harm and, where appropriate, to discuss the question of compensation.*

8.1. *Watercourse States shall cooperate on the basis of sovereign equality, territorial integrity, mutual benefit and good faith in order to attain optimal utilization and adequate protection of an international watercourse.*

8.2. *In determining the manner of such cooperation, watercourse States may consider the establishment of joint mechanisms or commissions, as deemed necessary by them, to facilitate cooperation on relevant measures and procedures in the light of experience gained through cooperation in existing joint mechanisms and commissions in various regions.*

Article 7 turned out to be the convention's most controversial provision. But it also proved to be inspiring. In its 1997 historical judgement on the famous Gabcikovo-Nagymaros case, the International Court of Justice referred to it implicitly, even though the convention was not binding yet (McCaffrey 2008). Hungary (the downstream party) complained that Slovakia (the upstream party) had undertaken some controversial hydro-engineering works once Hungary abandoned its contribution to a project that both countries agreed to in 1977. The Hungarian decision was motivated by environmental concerns, since the dam was likely to damage the Danube ecosystems. The court ruled that Slovakia and Hungary broke their agreement, but the original plan was indeed environmentally harmful, and hence it should be revised. It was remarkable that the court referred neither to the Danube conventions, nor to the Helsinki convention, to which both countries were parties.

Upstream and downstream countries' interests do not coincide. Consequently, how to reconcile their interests and how to achieve economic efficiency has been a challenge. The international community has developed a number of instruments to facilitate cooperation, but sovereign states tend to attach priority to their individual benefits even though – following adequate distribution – cooperation can enhance them.

2. Regional cooperation

2.1 Acid rain in Europe

This is no longer a burning issue in Europe. Yet several decades ago, it was widely debated by environmentalists and policy makers. In 1979 the Long Range Transboundary Air Pollution (LRTAP) convention was adopted by the UNECE. Thanks to the LRTAP, acid rain in Europe has been largely eliminated.

What is commonly referred to as "acid rain" is, in fact, a broad phenomenon. Any form of precipitation – not only rain, but also snow, fog etc. – may have an acidic reaction. Present in the atmosphere, water vapour – before it is deposited in the soil – can be acidified by certain chemical substances. Even if the earth atmosphere was free from any anthropogenic contamination, the reaction of the rain would have been slightly acidic. Carbon dioxide, i.e. the precursor of carbonic acid, is responsible for this. Acidity (or its reversal, i.e. alkalinity) is commonly measured in pH units. Pure (distilled) water has a pH measure close to 7, half a way between a pure acid (0) and a pure alkaline substance (14). Rainwater has a pH measure around 6.5 due to a light addition of carbon dioxide. A problem starts when the measure goes further down, for instance due to the addition of sulphur dioxide (the precursor of sulphurous acid) or nitrogen oxides.

Rain in Europe used to have a pH measure of 5 or even less. The record breaking rain had a pH lower than 3, just like lemon juice. The acidity of precipitation had an adverse impact on nature and human health. Forests and lakes were dying, and people suffered from upper respiratory diseases. Apart from that, economies suffered directly as well, since acid rain accelerated the corrosion of buildings and infrastructure in general.

The Swedes and Norwegians were concerned about acid rain earlier than others. Initially they failed to inspire wider international community. Despite efforts, the first of the so-called Earth Summits, organized by the United Nations in Stockholm in 1972, did not adopt any agreement on acid rain. However, by the end of the decade, a convention was signed to solve the problem permanently.

One may ask why the Swedes and Norwegians were among the first to struggle with acid rain. Perhaps sociologists would point to a high level of education and environmental awareness. But perhaps an even more important factor was the geological structure of the Scandinavian Peninsula. Thousands of years ago, glaciers pushed almost all the soil (rich in calcium) to the south, leaving the northern part of the continent with barren granite rock. Consequently acid rain in Poland does not lead to such catastrophic effects in ecosystems as in Scandinavia. The solution of sulphuric acid which comes with the rain reacts with calcites and loses some of its deadly power. As the Polish soils are more abundant with calcites than the Scandinavian ones, the effects of acid rain were more evident in Sweden and Norway.

Sceptics suggest that sulphur is a natural component of the biosphere, so attempts to eliminate it lack scientific justification. Indeed, sulphur is emitted from volcanoes and it is released as a result of rock decay, though the scale of these processes is smaller than anthropogenic emissions. At the turn of 1970s and 1980s, it was estimated that natural systems could absorb as much as 200 million tons of sulphur dioxide per year (Varhelyi, 1985). Natural emission is just above 100 million tons. At the same time, anthropogenic emissions reached almost 150 million tons per year, i.e. they definitely exceeded what can be tolerated by nature (Smith et al., 2010).

More importantly, almost half of anthropogenic emissions originated in Europe. Thus the continent became the most acidified pocket on the planet. The situation recently changed, and Latin American and Asian regions are at the top the list now.

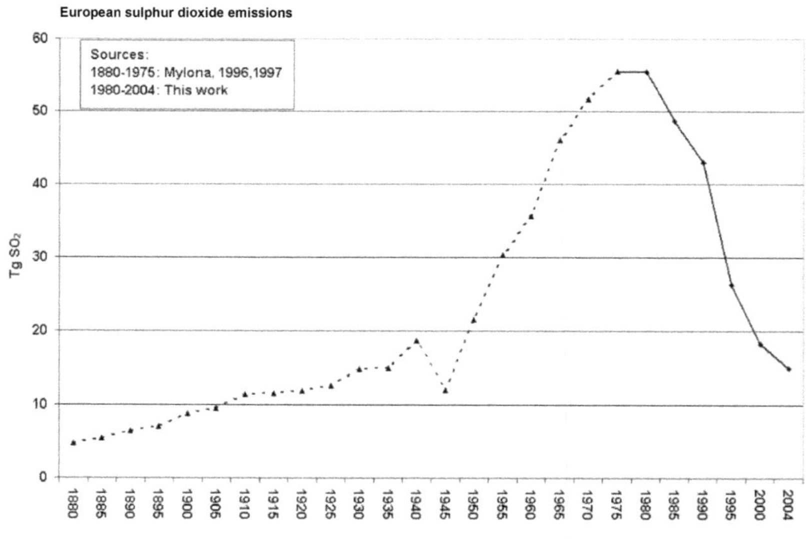

Source: Vestreng et al. (2007) (and Mylona (1996),(1997))

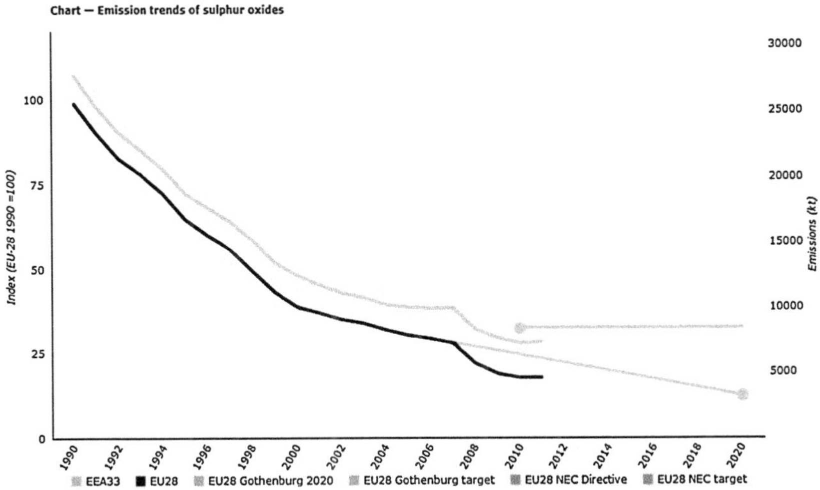

Source: EEA 2014

Anthropogenic sulphur emissions are caused mostly by fossil fuel combustion. Extracted from many deposits, hard coal is 1% (or more) contaminated with pyrite (a mineral containing sulphur). When this coal is burned, the pyrite is burned along with it, and sulphur dioxide is emitted. Liquid fuels are also contaminated. Crude oil contains sulphur compounds, so its combustion implies the emission of sulphur dioxide as well.

Acid rain has also been caused by nitrogen oxide emission, though the mechanism is somewhat different. Fossil fuels may be not contaminated with nitrogen compounds at all. Nevertheless ambient air is composed mainly of nitrogen. Its concentration is almost 80% (the rest, i.e. 20%, is filled mainly by oxygen). Under normal temperature, nitrogen does not react with oxygen. Increasing the temperature by several hundred degrees Celsius triggers chemical reactions. Burning fossil fuels causes the temperature to rise which leads to the emergence of nitrogen oxides. Not only power plants are responsible; car engines also emit nitrogen oxides. Indeed, today they are the key source of acidifying substances. In the 1970s – when the car fleet was much smaller – sulphur was considered the main culprit.

2.1.1 The Thirty Percent Club

Swedish and Norwegian efforts failed to bring a solution in Stockholm, but – continuing throughout the 1970s – they finally succeeded in 1979, when the Long Range Transboundary Air Pollution convention was signed in Geneva.

The convention was criticized and even ridiculed by some environmentalists as toothless, and the fact is that the convention did not commit its signatories to do anything substantial. It merely identified the problem and concluded that signatories should solve it. It is often the case that parties to negotiations may be so politically committed to reaching an agreement that they opt for a meaningless treaty in order to conceal their failure to negotiate an arrangement offering a more permanent solution (Sjöstedt & Svedin, 1993). Some critics saw the Convention as just such a failure. However, negotiating a "mild" convention that triggers on-going procedures for review and amendment is better than prospects for a stricter agreement postponed to an uncertain future (Sebenius, 1993). Indeed the Geneva convention included two important practical provisions. First, signatories are supposed to report their sulphur dioxide emissions. Second, they are supposed to meet annually in order to analyse the data.

The first of these provisions – seemingly quite innocent – was further softened by countries of the former Soviet bloc. The convention allowed signatories to substitute the emission reporting with reports on transborder fluxes. Consequently, East Germany did not report its total emission, but only its (approximate) part which crosses borders.

The *toothlessness* of the convention facilitated its universal acceptance. Almost all members of the UNECE ratified the convention (except for Romania and Yugoslavia; the former did it finally in 1991). More importantly, emission data began to be analysed. Some governments submitted inaccurate information (for instance, Poland reported emissions of 2.5 million tonnes, even though the true number was 4 million tonnes), but even such flawed statistics enabled the construction and analysis of simulation models linking the emission to the acidity of precipitation. After a couple years, acid rain emerged as an important theme of policy debates in Europe.

Despite its *toothlessness*, the convention started to work. Emission data was supposed to be analysed in Oslo, but the Soviet Union announced that it had the necessary computing potential as well. As a result, two simulation models were created: one in the Meteorological Synthetising Centre West in Oslo, and another in the Meteorological Synthetising Centre East in Moscow. From a purely economic point of view, this was waste of money. But from a political point of view it was beneficial, as it involved experts from both sides of the iron curtain.

Computer simulation models applied meteorological science in order to explain sulphur concentrations observed in several dozens of monitoring stations (there were two such stations in Poland) in terms of emissions from various regions in

Europe. It was clear that some of the emission data was false. Nevertheless, such false information was used. Among other things, the models included the variable "un-identified sources" (which could have been identified sometimes) to explain actually observed concentrations. For some countries these "un-identified sources" accounted for 30% of depositions, but this only called for more inquiries.

Over time the models improved, and by the mid 1980s nobody could question the fact that sulphur dioxide migrates all over Europe. Acid rain victims are not necessarily guilty for the pollution. Moreover, the models demonstrated that a significant part of sulphur depositions does not originate from the most annoying neighbour, but rather from several other countries. Otherwise, that is if an acid rain victim could identify a single country as the source of its problem, the solution could have been arrived at in bilateral negotiations. But reality turned out to be different. Simulation models convinced that solution can be arrived at in a multilateral agreement only. Perhaps this is the most important early success of the Geneva Convention.

Starting in the early 1980s, many tables characterizing European migrations of sulphur dioxide were compiled. They had the form of square matrices whose columns and rows included all the countries of the continent (usually with Turkey and the entire Soviet Union). Sometimes the Soviet Union was divided into its European part and the rest. The numbers were based on computations of the Oslo and Moscow Centres, which – by the way – were quite consistent. The top row of such a table listed emissions from all the countries concerned, while the last column – depositions in respective countries. The table's interior included transfers from an emitter to a country where the pollution was to be deposited. The table compiled at the University of Warsaw was based on the official convention's reporting with one exception. The official Polish emission was 2.5 million tonnes, whereas the actual one was 4 million tonnes; thus there was a need to allocate the missing 1.5 million tonnes to migrations originating from Poland. The table below is based on a report published at the University of Warsaw (Zylicz, 1985) and (Highton and Chadwick, 1982).

The table contains information on emissions originating from all European countries, and depositions located there (the first two columns). Then, depositions are related to countries' areas and translated into average annual concentrations (columns three and four). The next two columns include information on the share of imported depositions and the "Most Annoying Neighbour", i.e. the country whose emissions have the highest share in imported depositions. The last two columns include the share of depositions originating from the "Most Annoying Neighbour" and the share of "un-identified" sources (which could have been identified sometimes, but Synthesizing Centres did not want to question official reports).

Even though it contains outdated information, studying the table can be fascinating. According to professionals, mean average concentration of sulphur dioxide higher than 20 μg/m³ damages forests, and higher than 40 μg/m³ harms people. Hence the table shows that forests were threatened almost everywhere, and in some cases, human health was also affected country wide (excessive concentrations in smaller regions were certainly more frequent). But perhaps the most important message from the table is included in the second to last column. With some exceptions, depositions attributed to the "Most Annoying Neighbour" were lower than 15%. In other words, almost no country could solve the problem in a bilateral agreement with its "Most Annoying Neighbour". The solution must be a multilateral one including the entire continent.

Country	**Emission 1000 t**	**Deposition 1000 t**	**Density kg/ha**	**Concentr. μg/m³**	**Import %**	**M.A.N.***	**Share of M.A.N. %**	**Un-ident. %**
Austria, A	480	746	89	27	88	I	14	10
Belgium, B	850	397	130	39	47	F	14	5
Bulgaria, BG	880	741	67	20	59	RO	15	9
Czechoslovakia, CS	3610	3274	256	77	59	DDR	14	7
Denmark, DK	480	250	58	17	53	DDR	12	10
Finland, SF	600	962	29	9	71	SU	30	14
France, F	2870	2640	48	14	54	GB	9	15
East Germany, DDR	4220	1836	170	51	36	D	10	3
West Germany, D	3710	2614	105	32	53	DDR	11	7
Greece, GR	390	554	42	13	72	CS	17	15
Hungary, H	1840	1009	108	33	52	CS	13	5
Ireland, IRL	300	208	30	9	46	GB	14	30
Italy, I	3240	2093	69	21	32	F	5	11
Luxemburg, L	30	14	54	16	71	F	43	-

Country	Emission 1000 t	Deposition 1000 t	Density kg/ha	Concentr. µg/m³	Import %	M.A.N.*	Share of M.A.N. %	Un-ident. %
Netherlands, NL	510	391	95	28	73	D	25	7
Norway, N	150	637	20	6	90	GB	15	28
Poland, PL	4000	3539	113	34	54	DDR	14	5
Portugal, P	160	209	23	7	61	E	28	30
Romania, RO	2220	2103	89	27	57	SU	15	7
Spain, E	2330	1737	34	10	28	F	4	15
Sweden, S	490	1244	28	8	83	SU	13	22
Switzerland, CH	130	268	65	20	87	I	32	15
Turkey, TR	780	990	13	4	56	SU	12	19
USRR-Europe, SU-E	19940	16106	30	9	35	PL	7	2
USSR, SU	27550	.	.	.	.	.	.	.
Great Britain, GB	4540	2058	84	25	20	F	3	8
Yugoslavia, YU	960	1596	62	19	78	I	15	13
Total	**67320**	.	.	.	.	.	.	.

* M.A.N. – The Most Annoying Neighbour (the country where the highest share of "imported" depositions comes from)

The Geneva Convention is a successful example of what such an agreement should look like. It launched cooperation that was not spectacular in the beginning. On the contrary, it was criticized as "toothless". But it was founded on the universal reporting obligation, and gradually – through supplementary protocols – it acquired "teeth"; that is, it developed instruments to commit signatories to seriously address the problem.

Signed in Geneva in 1984, the first protocol to the convention was not spectacular either. It could have created an impression of a bureaucratic step, as it launched a research institution called EMEP (Cooperative Programme for Monitoring and Evaluation of the Long-range Transmission of Air Pollutants in

Europe). In reality, EMEP turned out to be a successful endeavour that led to a better understanding of European acid rain.

The second protocol, signed in Helsinki in 1985, represented a spectacular breakthrough. It has been called "The Sulphur Protocol" or "The Thirty Percent Protocol", as it obliged the signatories to reduce sulphur dioxide emission by 30% (with respect to 1980 levels) by 1994.

Environmentalists pointed out that – at last – the convention acquired "teeth", i.e. it started to "bite". In other words, signatories did not confine themselves to computing and analysing; they started to work instead. This, however, came at the price of losing the support of some parties. As many as 10 countries did not ratify the protocol. Poland – as well as Great Britain and USA – was among those who did not ratify. The American government stated arrogantly that it would fight acid rain at a rate and by dates that it deemed appropriate. In contrast, the British government explained politely that it could not accept the arbitrariness of the year 1980 as the basis for calculating the 30% reduction; it argued that, had 1975 been chosen as the base year, Great Britain would have complied with its 30% already. British emissions started to decline earlier (when newly discovered hydrocarbon deposits allowed reductions in coal combustion). The government of Poland justified its refusal on high abatement costs.

Some commentators observed that, for the first time since the end of the World War II, Poland allied with USA and Great Britain against Germany. Not only West Germany but also East Germany ratified the protocol, as did Bulgaria, Czechoslovakia, Hungary, and even the Soviet Union. Many countries from the former Soviet bloc signed and ratified the protocol in bad faith, anticipating that – in light of economic considerations – its requirements would not be fulfilled. Despite this, they chose such a solution for short-term tactical reasons, given that they did not want to trigger comments on excessive energy intensity in centrally planned economies.

Actually, an even more ambitious target of 40% (on average) could have been contemplated as the one which maximizes total regional benefits. However, as Mäler (1990) points out, the four net losers of such an agreement – the UK, Finland, Spain, and Italy – were better off staying out of it and "free-riding" on abatement efforts undertaken by the rest of the region.

Historical developments let Poland comply with the Helsinki Protocol with a large safety margin. After the collapse of the central planning system, economies of the former Soviet bloc reduced their emissions considerably thanks to the fact that the least efficient, energy intensive plants were closed. What was impossible to accomplish within the old system, for economic reasons, came as a side effect of the transformation process.

The Helsinki Protocol turned out to be more effective than expected, but indeed it was fairly arbitrary. First of all, the choice of the base year was arbitrary. Every country – like Great Britain – could insist that 30% is calculated from a different moment. Apparently the authors of the protocol concluded that any reference point would be contested by somebody, and something should be tabled. But the choice of 30% was arbitrary. There were no compelling economic arguments that the abatement target chosen maximized the difference between benefits and costs by 1994. Perhaps a more ambitious target would have been better, or – on the contrary – the target should have been more modest.

The protocol's largest deficiency is the principle of uniform reduction in all countries. It precludes achieving cost-effectiveness which calls that all participants face identical marginal abatement costs. Yet the requirement that everybody abates by 30% disregards any economic information. It implies that in some countries meeting the requirement costs more – per tonne of sulphur dioxide abated – and in some countries it costs less. Efficiency would have been improved if the protocol accepted that some countries comply jointly, or if the 30% were flexible: to let some countries do more, and some countries do less. Even though such a solution would have been better in terms of economics, it could turn out to be politically unacceptable. It is easier to sell an environmental idea emphasizing that everybody is supposed to do the same than arguing that everybody should face the same economic burden. Emissions can be measured in tonnes, by a universally accepted method. In contrast, measuring costs is subject to uncertainty.

The countries which signed the protocol were nicknamed "The Thirty percent Club". In the second half of the 1980s, there was a discussion in Poland about whether its membership in the Club was imaginable. Environmentalists argued that, even though abatement is costly, economic losses caused by acid rain are even higher. This discussion was purely academic, as it was not relevant for political decision making in a non-democratic system.

The collapse of the central planning system in 1989 allowed this problem to be solved. It became public knowledge that a part of heavy industry was operated not in order to contribute to economic surplus, but rather simply to sustain itself. Some analysts claim that energy consumption of some coal mines was higher than the energy that could be derived from the coal extracted. When plants started to calculate costs, such activity lost its *raison d'être*. Consequently the emission of various pollutants – including sulphur dioxide – decreased.

As a result of historical processes, *de facto* membership in "The Thirty Percent Club" was achieved by countries which did not ratify the protocol (like Poland) or ratified it without realistic prospects for compliance (like Czechoslovakia). From the point of view of environmental protection, a breakthrough was achieved: sulphur dioxide emission decreased drastically. Unfortunately, rain in Europe is still somewhat acidic. However, unlike in the 1980s, the problem is not so much caused by industrial plants, but rather by the transport sector. Diesel oil is contaminated with sulphur. Standards for liquid fuels are getting stricter, but emissions are still high as a result of growing mobility. An additional problem comes to the fore. Initially the acidification of the environment was predominantly caused by sulphur dioxide. Now it results, to a larger extent, from nitrogen oxide emission. Nitrogen oxides accompany high combustion temperature, such as achieved in a car engine. Hence, despite the fact that more cars are equipped with catalytic converters, and their engines are more efficient, the problem has not disappeared.

2.1.2 Protocols to the Geneva Convention

The first two protocols paved the way for the success of the Geneva Convention. The first one provided for a solid reporting foundation, while the second one gave the convention "teeth": it forced its signatories to abate. Both of them serve as milestones in the process of improving air quality in Europe. Subsequent protocols have proven to be remarkable as well. These are:

- Sofia Protocol, 1988, on nitrogen oxides,
- Geneva Protocol, 1991, on Volatile Organic Compounds (VOC),
- Oslo Protocol, 1994 (The Second Sulphur Protocol),
- Aarhus Protocol, 1998, on heavy metals,
- Aarhus Protocol, 1998, on Persistent Organic Pollutants (POP),
- Gothenburg Protocol, 1999, on acidification, eutrophication, and tropospheric ozone

All of them are related to air pollutants that migrate over large distances. This does not solve the air quality problem entirely. It has been universally known that ambient air quality depends also on particulate matter emission. However, particulates, especially those with a large diameter, are usually deposited close to where they were emitted. Hence solving the problem effectively can be done by local authorities. The protocols – in accordance with the scope of the LRTAP convention – deal with pollutants that travel long distances.

Each of the protocols addresses an important environmental problem and each deserves detailed analysis. Here only the Second Sulphur Protocol will be analysed, as it builds on the Helsinki Protocol, which was analysed in some detail above.

The Helsinki Protocol allowed for a struggle to be waged against European acid rain. It led to a significant abatement of sulphur dioxide by more than the agreed 30%. The success resulted in part from commitments envisaged by the protocol, and in part from historical processes caused by the collapse of the central planning system. In effect, European emissions in 1994 were even lower than 70% of the base level of 1980.

Nevertheless, environmentalists argued that European ecosystems were still excessively exposed to acidifying substances. What was needed in order to help the forests and reduce upper respiratory diseases was a 50% reduction rather than 30%. This called for a more ambitious target than adopted by the Helsinki Protocol, which was now called "The First Sulphur Protocol".

The Oslo Protocol, i.e. "The Second Sulphur Protocol" replicated certain provisions from its predecessor. As indicated above, despite its noteworthiness, the Helsinki Protocol was not free from arbitrary solutions. The first arbitrary element – the UK government protested against it – was the choice of the base year of 1980. The Second Sulphur Protocol arbitrarily adopted the same base year. This was perhaps a good solution, since opening up a discussion on alternative baselines would have not led to a universally acceptable solution. More importantly, the British government, which rejected it nine years earlier, accepted it now. Another arbitrary element was choosing the overall reduction rate equal to 50%.

However, "The Second Sulphur Protocol" left the rigid principle of uniform reduction in every country. The overall target of 50% emission reduction in Europe was allocated among the signatories, taking into account so-called *critical loads*, i.e. the maximum sulphur depositions that can be placed in certain regions without causing significant environmental damage. Given its geological structure, a typical hectare in Central Europe can absorb more sulphur than a hectare in the Scandinavian Peninsula. Consequently, the emission that – according to mathematical models of the European migrations of pollution – is likely to cause a higher deposition in an area regarded as sensitive should be abated to a larger extent. Eliminating all the violations of critical loads would be difficult. Therefore the protocol obliged its signatories to limit their emissions to such an extent that the number of violations is low.

The non-binding Annex I to the Second Sulphur Protocol includes a map of Europe disaggregated into several hundred squares (Poland is covered by 20 such squares) with critical loads per square meter. In Poland critical loads are estimated at roughly 1 g/m^2; in some areas in Southern Europe they are twice as high, while in Sweden and Norway they are many times lower. The idea behind the agreement is to gradually achieve emission reductions such that depositions observed in Europe do not exceed critical loads.

According to the Second Sulphur Protocol, Poland was to reduce its emission by 66% by 2010. The largest reduction (80% by 2000) was declared by Austria, Denmark, Finland and Sweden. Norway adopted a 76% target, and Germany – 83%. The latter was relatively easy, since the base included an enormous emission from East Germany which was vastly reduced just after unification. On the other hand, some countries adopted more modest targets. Bulgaria was supposed to reduce its emission by 45%, and Russia by 40%. Several countries (especially small ones) adopted extremely low targets. For instance Greece, Portugal and Spain – which did not comply with provisions of the Helsinki Protocol – agreed to very low abatement targets.

Critical sulphur deposition

(5-percentile in centigrams of sulphur per square metre per year)

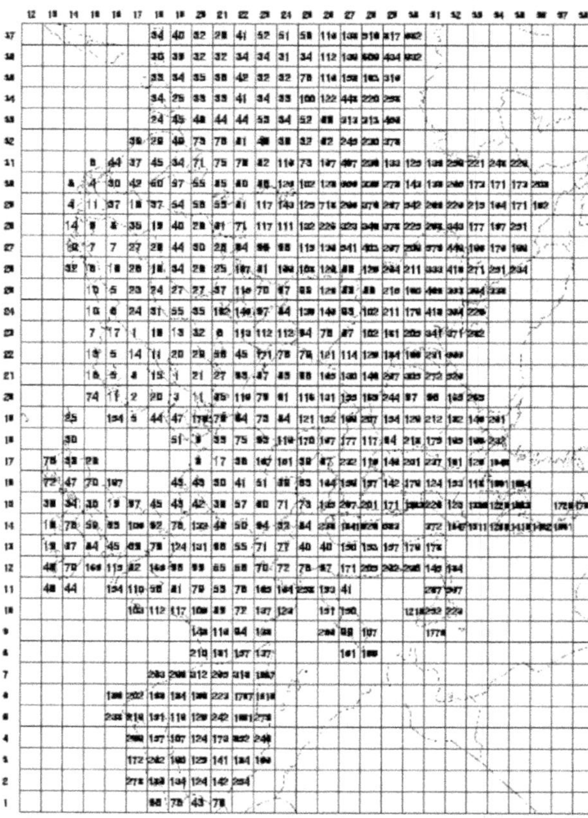

Annex I to the Second Sulphur Protocol

Poland did not ratify the Second Sulphur Protocol. Initially the Polish government was afraid that such a deep emission reduction would be too expensive and would harm economic growth. Research carried out at the University of Warsaw (Kiuila, 2000) did not confirm these concerns. On the contrary, it was estimated that the economic cost of ratification would be rather low, while domestic (mainly health) benefits would be high. In the meantime, emissions decreased to 974 thousand tonnes, i.e. much below what was required by the protocol. Yet the protocol includes a binding Annex V with desulphurization standards, and standards for sulphur content in fuels. These are more difficult to comply with. Even though the standards were accepted by the European Commission (in 1998) and included in its directives, Poland can request derogations. Thus, for the time being, the lack of compliance with all the details of the Second Sulphur Protocol would make its ratification difficult. On the other hand, as a member of the European Union, Poland is *de facto* bound by it anyway.

2.1.3 Non-European acid rain

The European Acid Rain problem has been largely solved. In the 21st century environmentalists point to East Asia and Latin America as regions most plagued by what was once considered a major issue in Europe. Even though the number of countries affected is probably lower there, the transboundary aspect of the problem is present as well. The Geneva Convention is thus worth studying.

North American acid rain was largely solved too. It was not as acute as in Europe, and probably easier to solve, given that there were only two major countries involved: the USA and Canada. Yet there were additional differences. In Europe there was a much stronger awareness of forest decline (Shaw, 1993). In contrast, in North America, acid rain was predominantly linked to the collapse of fishing in lakes. Therefore public support for anti sulphur measures was much more widespread in Europe. Moreover, in the USA and Canada governments disagreed over how to treat uncertainty in scientific evidence; US negotiators seemed to be more sceptical about the scientific basis for the underlying simulation models.

Even though there were only two major countries involved, cooperation was complicated because of their federal structure (states in the USA and provinces in Canada). Both countries revealed rather complex patterns of externalities and political competencies. Occasionally the cooperation applied instruments characteristic for domestic policies. For instance, in the absence of an international agreement, some of the worst polluters in the Canadian province of Ontario were forced by the Canadian federal government to decrease their sulphur dioxide emissions (Environment Canada, 1988). Of course, this could not solve the lake acidification problem, but it was a step towards a more permanent solution.

2.2 Baltic Sea eutrophication

2.2.1 The Baltic Sea predicament

The Baltic is perhaps the best researched sea in the world. Only the great American lakes (Ontario, Erie, Huron, Michigan and Superior) – whose transport role is sometimes considered equivalent to that of the Baltic Sea – have a similarly high number of scientific analyses per square kilometre. A great deal of information about the Baltic Sea is available. The distribution of temperature has been researched thoroughly. Likewise – the salinity, currents, and characteristics of habitats that can be found there. Unfortunately, a fairly good recognition of problems does not translate into protecting the sea.

As late as the 1950s the sea was fairly clean, even though many beaches were degraded, and numerous industrial plants discharged their sewage without any treatment. Problems were local, and they concentrated close to large harbour cities. Most of the sea ecosystems enjoyed good quality.

Serious problems emerged in the second half of the 20th century. Pollution loads kept growing, and the sea could not dilute them sufficiently fast. To some extent the functioning of natural processes is based on dilution. Various substances circulate and sustain living creatures. The chemical composition of the environment fluctuates, but as long as the concentrations are modest, even toxic substances may serve the development of some organisms. If their concentrations become excessive, the environment ceases to manage them. This is precisely what happened in the Baltic Sea in the middle of the 20th century.

Map of the Baltic Sea drainage basin (with sub-basins).
Source: Helcom (2011)

The Baltic Sea is unique. Elementary geography reminds us that ¾ of the earth surface is covered with water, and only ¼ with land. On average, three units of sea surface correspond to one unit of land surface. At the same time, the Baltic Sea area is 377 thousand square kilometres, and its drainage basin is 1642 thousand square kilometres. The proportion is thus reversed: one unit of sea surface corresponds to more than four units of the drainage basin. If one adds the low average depth – only 55 metres – what happens in the sea is determined by what happens on the land.

Other natural conditions make this predicament even more problematic. The sea contains about 22 thousand cubic kilometres of water. The exchange with the ocean is difficult, as it goes through the Danish Straits. The average annual outflow is roughly 940 cubic kilometres, which corresponds to an average exchange rate of 25 years. In the opposite direction – mainly as a result of winter storms (which do not occur frequently) – transfers from the North Sea are much lower. In an average year, rivers discharge some 660 cubic kilometres of fresh water. To balance the inflows and outflows one needs to take into account precipitation and evaporation. Because of separation from the Atlantic Ocean, the Baltic Sea is considered a semi-enclosed brackish sea. Exchange with the wider environment is difficult, and what is discharged from the drainage basin decides upon the chemical composition of the sea water.

Many people are passionate about pollution emitted from large industrial plants located close to the shore, or coastal cities not equipped with adequate sewage treatment plants. They are also passionate about news on Russian nuclear submarines sunk. Or about dumping in the Baltic Sea large inventories of Prussian chemical weapons during the World War I. All of these have threatened the Baltic nature, but – fortunately – they have not caused any major disaster so far.

Eutrophication is something that damages the Baltic Sea ecosystems permanently. This strange name means excessive fertilization. More specifically, scientists look at balances of elements – nitrogen and phosphorus – that they call "nutrients". Strictly speaking, neither the former nor the latter are contaminants. Both nitrogen and phosphorus are essential components of fertilizers applied in agriculture. The problem starts when their concentration exceeds what an ecosystem can absorb.

An oligotrophic aquifer is the opposite of a eutrophic one. In an oligotrophic aquifer, the availability of nutrients is low, which corresponds to low biological production and high transparency of water. For many years the Baltic Sea was oligotrophic, when concentrations of nitrogen and phosphorus compounds were low. The situation changed in the 20th century, especially in its second half.

A large quantity of nitrogen is available in the atmosphere. Here, however, it is chemically inactive. Chemically active nitrogen compounds are produced by the economy. Sooner or later they are washed out from the earth surface or rinsed off the soil and eventually reach the sea. Phosphorus is also a natural component of the biosphere, but human activities may lead to its concentrations that local ecosystems find excessive.

According to measurements carried out in the Baltic region at the turn of the 1980s and the 1990s, the annual inflow of nitrogen was around 900 thousand tonnes, and that of phosphorus – 40 thousand tonnes. Both now have moderate downward trends. The 2008 estimates (the most recent year with reliable estimates (Helcom, 2011)) are 700 thousand tonnes for nitrogen and 30 thousand tonnes for phosphorus. It would be wrong, however, to infer from this that nitrogen is the more abundant nutrient, and phosphorus – the less abundant one. In order to grow, sea plants require both nutrients in a proportion defined by the so-called Justus Liebig law. This law – discovered in the 19th century – states that living organisms use nutrients in certain fixed proportion. If one of the nutrients is relatively less abundant than the other, it is called "the limiting factor". Even if the other one is available in a higher quantity, growth will not be faster.

Nitrogen is considered the limiting factor in Baltic Sea ecosystems. In other words, phosphorus is so abundant that its further growth would not lead to higher biological production, while its decline would not lead to lower production. The level of production is determined by the availability of nitrogen.

The concentration of nutrients, especially nitrogen, determines plant production in the Baltic Sea. More specifically, it determines the production of algae. These are consumed by animals, and so on. The trophic chain of the Baltic Sea ecosystems does not have a sufficient capacity to eat all the algae. Those that are not eaten die naturally, and their remnants decompose, using up the oxygen dissolved in water; they spoil its transparency, and emit toxic substances such as hydrogen sulphide.

Baltic Sea eutrophication has been caused by the excessive inflow of nitrogen and phosphorus from the land. Hence, if the sea is to be protected, land based activities need to be controlled.

2.2.2 Gdansk and Helsinki Conventions

The predicament of the Baltic Sea has been an object of concern for several decades at least. Many people are upset by the closed beaches or murky coastal water. But some people are interested in more remote areas that cannot be seen from the coast. The condition of the sea is disturbing indeed. Because of eutrophication, the balance between plant and animal production was disrupted. The production of algae is excessive. Not eaten individuals decompose, use oxygen which is not available for other organisms, and the circle closes.

The problem will not be solved by taking better care of beaches. Toxic residue found on the shore can originate far in the open sea. The condition of the sea depends on what happens in the entire drainage basin. Thanks to long term research work, amounts of phosphorus and nitrogen – the main two nutrients responsible for eutrophication – can be analysed in detail. Their highest concentrations are found near the mouth of the Vistula river, in the Gulf of Riga, and in the Gulf of Finland. However, the circulation of water makes every part of the sea dependent on what happens everywhere else.

Therefore no country can solve the problem individually. Even worse, solving a local problem may harm other regions. The most drastic example is provided by a hypothetical clean up of the Gulf of Riga, which is presently highly contaminated by excessive nitrogen and phosphorus. Plant production is much higher than animals can eat. As a consequence, the aquifer works as a gigantic kettle where the two nutrients flow in an almost closed circuit. They are a nuisance for Latvians, but they "leak" into the Baltic proper to a limited extent only. If the Gulf of Riga were cleaned up, then – according to simulations – the outflow of nitrogen to the Baltic proper would increase since the local ecosystems would take it up to a lower extent. Since nitrogen is the limiting factor in the Baltic proper, algae blooms would become more frequent there.

The example above cannot serve as an argument that the Gulf of Riga should be "sacrificed" for the sake of the rest of the sea. It merely shows that the Baltic Sea should be looked at holistically, as its protection calls for activities that need to be coordinated and undertaken on a scale that does not have to appear justified locally.

Efforts to protect the Baltic Sea have a long history. In 1973 the Gdansk Convention on protecting living resources in the Baltic Sea was signed. An International Baltic Sea Fishery Commission was established in Warsaw to set fishing quotas. It soon became apparent that the condition of the sea depends on land activities rather than harvests. In 1974 the Helsinki Convention on protecting the marine environment of the Baltic Sea was signed. A totally amended agreement was signed in 1992, which is sometimes called the new Helsinki Convention. Its executive body, Helcom (Helsinki Commission), initiates programmes aimed at improving the overall quality of the sea. Contrary to the Gdansk Convention, the Helsinki Convention addresses land-based activities that are crucial for the protection of the Baltic Sea.

Initiatives of the Helsinki Commission are aimed at the protection of the entire sea, but Convention signatories are sovereign entities. Thus, decisions must be taken unanimously. Consequently improvements progress very slowly, at a much slower pace than in the case of Geneva Convention on long range transboundary air pollution. Presumably this has been caused by the fact that effects of air pollution were more evident while the eutrophication effects are perceived by political constituencies as less harmful.

The lower effectiveness of the Helsinki Convention does not mean that we can point to no successes. First of all, the inflow of nutrients stopped growing. The improvement has not been spectacular, but it exists. Moreover, Helcom managed to significantly lower the discharge of many toxic substances. An aggregate measure of improvement is halving the concentration of dioxins in eggs of common guillemot (*Uria aalge*), lower contamination with DDT decay products, heavy metals, and other similar trends (Helcom, 2010).

Limiting the use of DDT and heavy metals is fairly easy to mandate, since every government is aware that these substances are directly harmful to citizens and nature within its territory. Nutrients, i.e. the eutrophicating substances, are a different story. In the economists' jargon, Baltic Sea protection is a public good. Such goods are characterized by the non-exclusion and non-rivalry principles. The latter implies that socially justified supply of such goods is higher than the quantity resulting from the preference of any individual user. At the same time – because of the former – no one volunteers to finance the cost of supply, knowing that nobody can be excluded from the set of beneficiaries (once the supply is financed by somebody else). Everybody would profit from the Baltic Sea protection, but nobody is keen on financing it, since perhaps somebody else will pay. As a result, signatories of the Helsinki Convention contemplate ever improved protection programmes, but they spend on these programmes less than would be necessary to reduce eutrophication effectively.

The problem is further complicated by the fact that abating nutrients (nitrogen and phosphorus) provides local benefits too (along their way to the sea), but domestic priorities are not necessarily identical with wider regional ones. For instance, from the Polish domestic point of view, the most urgent concern is abating sewage discharged into tributaries of the upper Vistula. For example, protecting the Dunajec river (a tributary to the upper Vistula) is certainly beneficial for the entire Baltic Sea, but it contributes to the success to a small extent only. At the same time, from the point of view of providing people with safe potable water, and guarding local ecosystems, the protection of the Dunajec river is of paramount importance. Consequently, if the Polish government were sensitive to local benefits, it would support building sewage treatment plants in the South of the country. Yet the nutrients discharged in upstream regions are partially retained along their way to the sea. Out of, say, 12 tonnes of nitrogen discharged into the Dunajec river, only 4 tonnes will reach the Baltic Sea. If the discharge were reduced by 75%, only 1 tonne would reach the sea. In other words, the discharge is reduced by 9 tonnes, and the Baltic Sea receives only 3 tonnes less. If a similar exercise were carried out for the lower Vistula, say, for the city of Tczew, where almost the entire load reaches the sea, the same abatement effort (discharge of 3 tonnes instead of 12 tonnes) lowers the inflow into the Baltic Sea by the full 9 tonnes. Thus, from the point of view of protecting the Baltic Sea, the most effective sewage treatment plants are those located in Northern Poland.

All signatories to the Helsinki Convention are aware of this. All of them protect water ecosystems, but they do it because of their domestic benefits. The Baltic Sea attracts a lot of attention, but most activities are undertaken in order to protect inland ecosystems rather than the sea.

Abatement costs of reaching a mandated level of improvement do not seem to be difficult to compute. And yet they are. It can be calculated how much should be spent on equipping all municipalities with biological sewage treatment plants, how much it will cost to build the necessary sewerage system, and how much an average household will have to pay per cubic meter of sewage. Those numbers are not determined "objectively", since they depend on the professionalism of municipal officers, geography, and population density, among other things. Nevertheless perfect knowledge can be assumed, and theoretical expenditures to meet expected municipal sanitation targets can be computed.

One may dispute whether in sparsely populated areas it is better to invest in a sewerage system, or rather tolerate individual septic tanks, some of which will be leaking. One may dispute whether it is better to pay inspectors who check these septic tanks, or to have an expensive sewerage system which is more cheat-proof. Finally, one may dispute whether to invest in a more expensive technology whose maintenance is cheaper, or to rely on a cheaper technology which requires higher operational expenditures. These are questions that any well-educated engineer can address.

A problem that transcends such engineering dilemmas is a choice of alternative methods to limit the load of nitrogen and phosphorus into the sea. From the point of view of reducing eutrophication, it is both effective to equip housing districts with adequate sewage infrastructure, and to limit car traffic, which is responsible for part of nitrogen compounds reaching the sea. Should a decreased inflow of phosphorus be caused by a lower consumption of detergents, or a lower application of fertilizers in agriculture? Is eutrophication to be reduced by relying on highly efficient tertiary installations instead of cheaper secondary ones, or by restoring coastal wetlands, which serve as "nutrient traps"?

These sorts of problems are analysed by economists who study the costs of reaching a given target – here reduced eutrophication of the Baltic Sea. They start with the cheapest option. Here it is restoring coastal wetlands which were once drained. But the amount of nitrogen and phosphorus that can be abated by this method is small. Therefore they take the next abatement method. Let us assume that this is reduced fertilization in agriculture. Applying this method is not free, as it implies lower crop yields, i.e. a loss that needs to be compensated somehow. Alternatively, it can be reached by investing in better storage of slurry.

But even if all farmers switched to a better technology, the eutrophication problem of the Baltic Sea would not be solved. Thus the next abatement method, say, secondary treatment, should be considered. Then tertiary treatment, and so on, until the expected degree of reduced eutrophication has been reached.

It is hence evident that the cost of the Baltic Sea protection depends on the ambition level. A moderate programme of reduced eutrophication is cheap. Further improvement requires growing unit costs. A 50% reduction of inflow of nitrogen and phosphorus is contemplated within the framework of the Helsinki Convention. The annual cost of such a programme was estimated at 4 billion dollars. Of course, it can be higher, if unnecessarily expensive measures are implemented – for instance reduced fertilizer application in agriculture before restoring coastal wetlands that deliver the same result at a lower cost; or applying tertiary treatment, even though secondary treatment would be sufficient in certain cases, and so on. Economists say that reaching the 50% eutrophication target requires 4 billion dollars per year if a cost-effective programme is implemented (otherwise it could be more expensive). In this computation, researchers concentrated on nitrogen abatement for two reasons. First, nitrogen was identified as the limiting factor of eutrophication; and, second, nitrogen abatement measures often abate phosphorus as well. The following table summarises an assessment of cost-effective abatement in the whole Baltic region.

Cost-effective abatement in the Baltic drainage basin

Reduction rate (%)	5	10	15	20	25	30	35	40	45	50
Cost (10^9 \$/year)	<0.1	0.1	0.1	0.2	0.3	0.5	0.8	1.2	1.9	4.1

Source: Gren *et al.* (1995).

Similar assessments were repeatedly carried out later on, most recently by Wulff *et al.* (2014) and Hasler *et al.* (2014). They are higher in nominal terms – more than 4 billion euro in 2011 – but given the time elapsed since the earlier estimations, in fact, they are somewhat lower in real terms. An important reason why these newer estimates suggest a lower amount is that – unlike before, when cost-effectiveness was to be achieved by differentiating targets for countries – here cost-effectiveness is to be achieved by differentiating targets for smaller regions. Moreover, the total abatement cost turns out lower, if the ambition level is lowered, as in recent Helsinki Commission documents. As well, they can be lower because of the lower baseline loads than those observed in the 1990s (Ahlvik *et al.*, 2014).

Having grasped a rough estimate of the cost of the Baltic Sea protection programme, economists started to analyse expected benefits. Here the problem turned out to be more difficult. Reduced eutrophication – contrary to e.g. reduced travel time – does not involve market transactions. Consequently, no prices can be observed to inform about benefits gained.

Over the last couple decades tremendous progress has been made with respect to assessing values of non-market goods. What seemed impossible to measure in the first half of the 20th century can now be calculated, even though it may require substantial effort, and may involve making controversial assumptions.

Whenever they cannot rely on market prices, economists look at so-called surrogate or hypothetical markets. The former are markets where goods related to the analysed object are traded. For instance, there is no market for silence, and consequently silence does not have a market price. However, there is a market for real estate, and if one compares two almost identical commodities except that the first is located in a noisy neighbourhood, and the second in a quiet one, then the second commodity has probably a higher price. The difference in prices can be attributed to silence and that is how economists try to measure it in an indirect way. A hypothetical market is where the analysed object is to be provided, if people are willing to pay for it, or where they are willing to accept compensation for being deprived of it. By looking at this *Willingness To Pay* (WTP) or *Willingness To Accept* (WTA), economists try to measure the value directly.

Both surrogate and hypothetical markets help economists measure what seemed to be impossible to quantify before. Of course, one can think of myriads of reasons why the numbers arrived at may be misleading. Nevertheless, if the valuation procedures comply with requirements that are being refined constantly, the numbers are not less reliable than those observed in real markets (Carson and Czajkowski, 2014).

A University of Warsaw team of environmental economists decided to meet a challenge in the 1990s. It looked for either surrogate or hypothetical markets to shed light on what benefits are likely to occur when the eutrophication of the Baltic Sea is reduced. For instance, it was analysed whether Baltic navigation could become more profitable. Or whether the owners of real estate located close to clean beaches would benefit from higher prices. Or whether fishermen harvesting in a cleaner sea would enjoy greater catches. Yet – for diverse reasons – analysing such surrogate markets could not provide adequate information on clean up programmes. Parenthetically, eutrophication can be beneficial for fisheries, since in an oligotrophic sea, the population of many species is lower.

Under such circumstances, economists rely on hypothetical markets, which boils down to interviewing people in order to check how much they would be willing to pay to get something. Coordinated research programmes were carried out in Poland, Sweden and Lithuania in the mid-1990s. The Polish programme was the most extensive. This allowed to consistently interpret the Swedish and Lithuanian data that otherwise would be difficult to compare. As a result of this research, it was estimated that annual benefit from the 50% reduction of the Baltic Sea eutrophication is \$56 for an average adult Pole, \$28 for a Lithuanian, and \$229 for a Swede (Markowska and Zylicz, 1999).

The differences are easy to interpret. The low number for Lithuania is obviously caused by the lower wealth of a Lithuanian household. In addition, people in Lithuania may also feel other needs whose weight is relatively higher. Consequently a "cleaner Baltic Sea" is perceived as less important. The high number for Sweden is obviously caused by the much higher welfare there. But this is not the end of the story. In Poland, the share of people who go to the sea coast and care for the sea is lower than in Sweden. An average Swede lives closer to the sea than an average Pole. And – more importantly – almost everybody has a summer cottage (being his or her property or the property of a friend) located in a coastal region. As a result, for an average Swede, the Baltic Sea is more precious than for an average Pole.

Following a complicated extrapolation procedure (Markowska and Zylicz, 1999), these average numbers allowed researchers to estimate the total benefits from the 50% abatement programme at \$6 billion per year. Newer and more precise estimates are somewhat lower, but difficult to compare, as they are based on different programmes (Ahtiainen *et al.*, 2014). They correspond to more recent official Helsinki Commission documents. The abatement cost computed in these documents is lower and the new benefit estimates are higher than the cost, thus demonstrating the efficiency of the targets adopted. In other words, these official programmes can be justified in economic terms. A practical implication of this assertion is problematic, since expenditures are justified only when one looks at the sea as a whole, i.e. as a public good.

2.2.3 Prospects for a regional policy

Estimated costs and benefits of the Baltic Sea suggest that the programme contemplated by environmentalists is economically justified. In other words, when the entire drainage basin is taken into account, the benefit from the 50% reduction of nutrient discharges is higher than its cost. But costs and benefits need to be compared at the drainage basin level. If balances were made at the level of individual countries, the 50% target would turn out to be too ambitious. The cost is the same as accounted for in the drainage-basin-wide case, but the benefit from improving the quality of local ecosystems is not attractive enough. Some of the benefits are external in the sense that they accrue to other countries.

The Baltic countries protect the water, but the scale of their activities is adjusted to domestic priorities. The Swedes do whatever they consider justified and complain that the Poles do too little. The Poles also do whatever they consider justified and declare that they could have done more if the beneficiaries participated in the expenditures. There seems to be no way out of this situation: there will be no progress beyond what everybody does according to their individual priorities, unless a mechanism is in place to finance additional efforts. Such efforts may prove necessary in order to protect the sea as a whole (public good, in economists' jargon), but they are not justified from the point of view of an individual country.

A particular mechanism for financing a public good was first suggested long ago, but it was formalized as late as the turn of the 1980s and 1990s. It is now called the Chander-Tulkens (1992) model, referring to the names of the economists who developed it using very sophisticated game theory techniques. Based on their theoretical model, one can derive the following equation to determine money transfers between the users of a common public good:

$$T_i = \gamma_i p_i - (\pi_i{:}\pi_N)\cdot\Sigma_j\, \gamma_j p_j,$$

where:

T_i – money transfer to country *i*,
γ_i – marginal abatement cost in country *i*,
p_i – pollution abatement in country *i*,
π_i – benefits in country *i* from the region-wide abatement,
π_N – the sum of benefits from the region-wide abatement
($\pi_N = \Sigma_j\, \pi_j$).

Irrespective of the complicated method to derive this formula, its intuitive interpretation is very simple. Every country gets its abatement cost financed ($\gamma_i p_i$), but at the same time it contributes to the total regional abatement cost ($\Sigma_j\, \gamma_j p_j$) in proportion to its share in total benefits ($\pi_i{:}\pi_N$). A negative amount means that a country pays rather than receives money. It can be easily demonstrated that the sum of transfers is zero ($\Sigma_j\, T_j=0$). This mechanism allows for the financing of the economically justified supply of the public good, while beneficiaries participate in this endeavour in proportion to the benefits enjoyed.

For the Baltic Sea clean up programme – understood here for simplicity with the 50% nitrogen abatement – π_N is estimated at $6 billion, and $\Sigma_j \gamma_j p_j$ is estimated at $4 billion. The Chander-Tulkens model lets us to calculate hypothetical money transfers allowing countries to reach such a state. The following table was compiled at the University of Warsaw.

Hypothetical Baltic transfers

Country (*i*)	$\pi_i{:}\pi_N$ [%]	T_i [10^6 $]
Finland	14.4	-216.9
Sweden	26.7	-395.6
Denmark	16.5	-292.3
Germany	11.2	67.2
Poland	24.1	280.8
Lithuania	1.2	280.0
Latvia	0.8	208.8
Estonia	0.6	177.2
Russia	4.6	-109.2
Total	100.0	0.0

Source: Markowska and Zylicz (1999)

As could have been anticipated, Sweden emerges as the largest net payer, and Poland – as the largest net beneficiary. Quite unexpectedly, Germany turns out to be financed in net terms, while Russia turns out to be a net payer. The paradox can be easily explained by the benefit extrapolation method applied. The benefits are assumed to be proportional to the number of people living in the Baltic drainage basin (in Poland it is virtually the entire population). The German benefits were extrapolated from the Swedish data, but the total is small given that only 3 million adult Germans live in the Baltic drainage basin. In particular the inhabitants of Berlin are excluded from the calculation (as they live in the North Sea drainage basin), but they surely benefit from the Baltic clean up programme. Thus if the German benefits were calculated more adequately, they would turn out probably higher than the 11.2% number from the table. For Russia, the benefits were extrapolated from the Lithuanian data which could have resulted in their overestimation (Russian share in total benefits is probably lower than 4.6%). Lithuanians are proud to be a coastal nation, they seem to appreciate the sea and – on average – presumably appreciate it more than Russians.

The hypothetical transfers described in the table are disputable and can certainly be corrected once more accurate knowledge on benefits from reduced Baltic eutrophication is available. Their order is remarkable though. They are likely to be over one billion dollars per year. This is many times more than Nordic countries allocate as environmental assistance for their Eastern and Southern neighbours to be spent on water protection. Therefore the largest net beneficiaries finance the reduced eutrophication of the Baltic Sea to a much lower extent than required in order to achieve an efficient scale. On the other hand, countries – such as Poland – that could have done more, keep doing what is determined by their domestic considerations. In order to abate nitrogen and phosphorus at the level justified by benefits accruing to all Baltic countries, a regional transfer mechanism should be established. Apparently signatories of the Helsinki Convention are not ready to take such a step.

The idea of a Baltic-wide fund has its predecessor in the idea of the European Sulphur Fund advanced by Mäler (1990), when prospects for abating acid rain were much less promising than what actually happened afterwards.

One reason that countries hesitate to establish a transfer mechanism is uncertainty as to their share in the total benefits. While average WTP for a given country is relatively easy to compute, an appropriate extrapolation poses a challenge. The method adopted in calculations summarised in the table above took into account the population living in the Baltic Sea drainage basin. Two littoral countries – Germany and Russia – have very small areas included in the drainage basin, which implies uncertainty as to how to account for their citizens who live in other drainage basins. In the case of Germany, extrapolation limited to those who live in the Baltic drainage basin led probably to a massive underestimation. On the other hand, the approach adopted in Ahtiainen *et al.* (2014) is probably also inappropriate. Their study assumed that populations of littoral states are homogeneous with respect to their WTP for the Baltic clean-up programme. Thus the German average WTP is multiplied by the entire German population. The table below summarises transfers implied by benefit shares based on Ahtiainen *et al.* (2014).

Alternative hypothetical Baltic transfers

Country (*i*)	$\pi_i{:}\pi_N$ [%]	T_i [10^6 \$]
Finland	4.2	206
Sweden	15.9	52
Denmark	3.5	245
Germany	47.3	-1428

Country (i)	$\pi_i:\pi_N$ [%]	T_i [10^6 $]
Poland	8.3	936
Lithuania	0.6	305
Latvia	0.3	228
Estonia	0.7	171
Russia	19.2	-715
Total	100.0	0

Source: Author's calculations based on Ahtiainen *et al.* (2014)

Transfers implied by this breakdown of benefits are completely different from the earlier pattern. First of all, the total sum of transfers is more than twice as high, which reflects the fact that the distribution of abatement costs and benefits are entirely different. German abatement costs are very low since a very small area is in the drainage basin. At the same time, the benefits are extremely large since the German population is much higher than that of any other Baltic country, except for Russia, which is assumed to be another large beneficiary (and net payer). Ahtiainen *et al.* (2014) take into account the population of Western Russia only, which nevertheless dominates over the rest. In addition, the table makes Germany and Russia net payers, and every other country – net beneficiaries.

The idea of taking entire countries instead of their parts included in the drainage basin is sound; hypothetical agreements would be signed by states rather than smaller territorial units. On the other hand, it is questionable whether extrapolation of the computed WTP onto the entire German population is justified. Extrapolation of the Russian WTP is also debatable. It is obvious that many Russians are aware of the Baltic Sea and do care about its environmental predicament. Yet it is doubtful whether they would agree to participate in a transfer mechanism leading to their subsidisation of the Baltic clean up in Finland, Sweden and Denmark.

Exercises based on alternative distributions of benefits from a less eutrophicated Baltic Sea demonstrate that hypothetical transfers are difficult to ascertain. They show that the more distribution of benefits differs from the distribution of abatement effort, the more extensive transfers are called for. If both distributions were the same there would be no need for transfers. Another lesson learnt is that transfers are sensitive to countries' WTP for the Baltic Sea clean up. Even a small change in the benefit assessment may result in a substantive change in hypothetical payments and in the moving of a country from the net payer to the net beneficiary group or *vice versa*. While computing individual WTP for the

Baltic Sea clean up seems to be a relatively easy task, extrapolating its results onto wider populations poses a substantial challenge.

Eutrophication is an example of a problem that cannot be tackled successfully by any single Baltic country, but it is not the only one. There are at least two others that have emerged as affecting adversely the entire Sea: dissolved organic matter (DOM), and micro-littering. The former is of interest not only as a marine pollution issue but also because of its potential impact on the global climate.

DOM includes particles that have a diameter lower than 0.00045 mm. They are not necessarily toxic, but they may add colours and consequently decrease water transparency. Given that the particles include carbon, they are of interest from the carbon cycle point of view. Their specific impact (both on the climate and on the Baltic Sea predicament) is not clear, as empirical data are still insufficient to draw firm conclusions (Vahtera *et al.* 2007).

In contrast, micro-littering can be easier to spot because the diameter of particles is less than 5 mm, but larger than 0.3 mm. There is thus a "grey" area of particles smaller than 0.3 mm, but larger 0.00045 mm. Even though they exist, they comply neither with the definition of DOM, nor micro-litter. They will probably be the subject of research in the future.

For the time being, micro-litter research is in its early stage. This much is known: micro-litter pollutes the sea and it can migrate over large distances. It originates from "ordinary" litter – such as packaging – which was broken into smaller pieces, or from the wear-and-tear of useful things – such as clothes. Hence, despite what many people deem, micro-litter does not have to be plastic; it can be "natural", like in the case of fibres from clothes torn in a washing machine. Moreover, it does not have to be toxic. However, micro-litter often serves as a vector for toxic compounds. As biological research demonstrates, cells can be poisoned by toxic compounds attached to particles that were eaten mistakenly by animal organisms. Even though the evidence of damages done by micro-litter in the Baltic Sea is still very limited (Magnusson, 2014), it is likely to draw increased attention in the future.

2.2.4 Implementation of hypothetical transfer mechanism

Hypothetical transfers based on the Chander-Tulkens model reflect side payments required to trigger cooperation between Baltic countries necessary to escape from the trap they are stuck in when they ignore the public good nature of the sea. They are rough estimates of what may provide incentives to cooperate, and they do not have to be understood as direct payments. Recent research confirms that free-riding is the most important obstacle to reducing eutrophication efficiently (Ahlvik and Pavlova, 2013).

One institutional design which may be contemplated is nutrient trading. As is widely known in environmental economics, tradable permits allow for the cost-effective allocation of an abatement effort. In other words, if the sum of permits is equal to what is agreed upon as the allowable level of discharges, their tradability will lead to minimising the total cost of reaching this level. Cost-effectiveness, however, is independent of burden sharing. In particular, if the permits initially allocated replicate the cost-effective apportionment of abatement targets (so that no trading will occur) there are no transfers. If a country receives more permits than implied by these targets, it will be a net beneficiary of trading. Otherwise it will be a net payer. Consequently, by manipulating initial allocation of permits, one can reach the transfer pattern identified by the Chander-Tulkens model.

An alternative way to implement the transfer mechanism is development assistance as practised by the Nordic countries. Within the framework of regional environmental aid, they have been involved in bilateral programmes to invest in abatement capacity in Southern Baltic states. As indicated above, the scale of these activities is at least one order of magnitude lower than the flows envisaged in the Chander-Tulkens model. In order to play an incentive role they should be increased many times, but the mechanism is present.

Yet another implementation of transfers – at least with respect to the European Union countries – could make use of the European structural and cohesion funds. Two problems need to be sorted out. One is their statutory function. They are supposed to narrow development disparities among regions and member states. At the same time, the purpose of Baltic cooperation is different, even though – to some extent – undertaking abatement projects can be seen as a development objective. The second is additionality both on the collection and the disbursement side. European fund revenues are built up through member states' contributions, which are motivated by certain needs articulated by the European Commission. Planned disbursements are carefully assessed by "cohesion" countries who negotiate their totals. Hence, if the Chander-Tulkens transfers were to be channelled by European funds, the additionality of their budgets has to be checked.

Finally, it should be noted that the idea of side payments is inconsistent with the *Polluter Pays Principle*. Some environmentalists object to them on these grounds and expect that all countries undertake whatever abatement activities are justified by supranational considerations. While consistent with some theoretical deliberations, the principle is not followed in international agreements. Even though it sounds odd, chapter 1 was devoted to the *Victim Pays Principle* as the one often applied in such agreements.

3. Protecting the Global Commons

As explained in chapter 2, delivering regional environmental cooperation is far from trivial even if the number of cooperating countries is small, and even if they seem to be close to each other both geographically and psychologically. Co-operation prospects become even more problematic when countries are more numerous, and more distant from each other. In this chapter we offer perspectives on how such cooperation may develop. It includes examples of success, as well as cases demonstrating that countries fail to deliver a global common good.

As before, analysis refers to the definition of a public good as one which is non-rival and non-exclusionary. Economic theory explains that the efficient supply of such a good complies with the following (so-called Samuelson) criterion:

$MC_i = \Sigma_i MB_i$,
where:
MC_i – the marginal cost of delivering the good incurred in country *i*,
MB_i – the marginal benefit from delivering the good for country *i*, and
summation extends over all the countries (using the good).

At the same time, unless cooperation is enforced, countries are likely to finance the supply of the good just to meet the following (so-called Nash) criterion:

$MC_i = MB_i$.

Cooperation can use the Chander-Tulkens model introduced in chapter 2, but countries first need to agree that they wish to overcome the inefficiency. Sometimes they do it, though sometimes they prefer to stick to an inefficient solution. By looking, in turns, at the ozone layer, climate protection, and biodiversity, we try to understand how cooperation develops.

3.1 Ozone layer

3.1.1 Ozone depletion story

We cannot see the thin ozone layer which cushions the Earth at the altitude of a dozen kilometres. Even physicists and chemists did not predict it. It was discovered as late as the 1960s thanks to measurements carried out in the course of space exploration programmes. At the same time, its beneficial role in protecting life was discovered. It is known that the Earth's atmosphere consists in 20% oxygen, whose compounds include two atoms of the element (O_2). Ozone, i.e.

three-atom oxygen (O_3), can be rarely spotted at the Earth's surface, and it is a harmful substance here. In contrast, in the upper atmosphere – so-called stratospheric ozone – it is not harmful at all, and it protects against excessive ultraviolet radiation. It creates a shield that controls the amount of this radiation and lets reach the Earth's surface only what is necessary for the people.

Excessive ultraviolet radiation has been identified as a factor which increases the risk of skin cancer. This is not its only role. Excessive ultraviolet radiation damages crops and speeds up the corrosion of building materials.

As soon as the ozone layer was discovered, it was measured in a systematic way. Its changes were detected, but – due to a software error – it took some time before analysts realized that "holes" in the ozone layer are not random. The lowest values for the width of the ozone layer were recorded in polar regions and in certain seasons, but the statistical programme which interpreted the data considered these *outliers*, and disregarded them. Later analysis of the raw data led scientists to discover the "hole" which is responsible for excessive ultraviolet radiation reaching the Earth's surface in certain areas systematically.

Looking for reasons started immediately. It turned out that, among other things, freons are responsible for damaging the stratospheric ozone layer. Freons – or chloro-fluoro-carbons, CFC – were discovered by the American concern Du Pont. They were patented in the first half of the 20th century, and were considered a miracle of industrial chemistry. They were chemically inactive, hence they behaved like perfect gases, but their production cost was significantly cheaper. Soon they became an indispensable component of industrial processes and products essential for contemporary civilization.

They were applied wherever a liquid (or a gas) was needed which did not react with other components or with the wall of a container. Until recently CFCs circulated in refrigerators. What is even more important, they were used in industrial cooling systems. CFC leakage from a household installation could happen only in the case of its malfunction or disposal. In industrial cooling systems it used to flow in circuits that were not quite closed; and hence its leakage into the atmosphere was routine.

CFCs were commonly applied as propellants in sprays, as they did not react with paints, cosmetics, and medicines kept in pressurized containers. They were applied in the electronic industry to clean microchips before assembling them in appliances such as computers, mobile phones, cameras, etc. CFCs were also used in the construction business, among other things in styrofoam production.

Widespread use of CFCs made governments reluctant to contemplate their ban. Nevertheless, they were more likely to be regulated than other areas of human activity that were also found responsible. It turned out that the emission of freon was not the only factor responsible for the ozone hole. Freon destroy the O_3 compounds chemically. Indeed they behave like perfect gases near the Earth surface. However, the conditions prevalent in the stratosphere – such as pressure and temperature – make them change their chemical characteristics. Still, the ozone layer can be destroyed not only chemically, but also physically. For instance, it is "punctured" by spaceships launched. Spaceships – placing communication satellites or many other objects – are launched almost every day. Ordinary passenger planes travel at the altitude of about 10 kilometres, i.e. they do not reach it. But supersonic jets – reaching the altitude of 20 kilometres – "puncture" the ozone layer twice: once when they acquire their cruising altitude, and then when they prepare for descent. There are hundreds of military supersonic jets taking off and landing every day.

Are there many people brave enough to ask for a space exploration programme to be scaled down or a military fleet reduced? From the beginning it was clear that the only realistic programme of ozone layer protection would attack freon alone.

Even though the ozone layer does not have a uniform width, and the "holes" can be geographically localized, from the economic theory point of view it is a global public good. Specific coordinates of CFC emission site are irrelevant. Once emitted, before it reaches the stratosphere, it is dispersed so quickly that it can attack ozone particles somewhere else. Therefore no country could effectively protect "its" ozone layer, as limiting its emission of CFCs does not translate into a lower CFC amount active over its territory. Moreover, a unilaterally adopted CFC limit – resulting in lower production of styrofoam, among other things – would provide incentives to produce this styrofoam elsewhere and would lead to CFCs being emitted elsewhere. Economists call this phenomenon *leakage* of the harmful substance. If the substance is only locally harmful, then the unilateral limit solves a local problem at the expense of deterioration somewhere else. But if the substance – like freon – is a global pollutant, then a unilaterally adopted constraint can be harmful for the region that introduced it.

It was thus clear that repairing the ozone "hole" requires setting a global limit for using freon. But the powerful Du Pont was the owner of the patents. Freon production brought handsome profits, and the company did not intend to lose this source of revenue. Consequently, for many years – especially at the turn of the 1970s and the 1980s – it put a lot of effort into campaigns discouraging the attack on freon. It argued that freon is an indispensable part of contemporary civilization. Without freon, a modern household could not function (can anybody think of living without a fridge?), but also industry is threatened (what would happen to the production of electronic devices or to thermal insulation?). Despite that, some countries introduced bans or constraints on using freon. The US banned some CFCs in "non-essential uses" in 1978, and European countries limited their production in 1980 (Haas, 1992; Széll, 1993). Of course, such unilateral steps had no impact on the ozone hole.

It is somewhat amusing that as late as the early 1990s one prominent Polish environmentalist warned publicly of the hunger catastrophe to be caused by the collapse of the refrigerator market expected as a result of reduced freon production.

In the early 1980s, efforts aimed at protecting the ozone layer boiled down to a ban on freon production. Proponents of the ban included environmental NGOs. Du Pont and institutions that supported it – including the US government – played the role of the principal opponent. Environmentalists seemed to be puzzled when in 1987 Du Pont changed its position suddenly, and started to demand the introduction of an immediate ban on freon (!).

3.1.2 DuPont *versus* ICI

Although Du Pont was the sole owner of CFC patents, it was not the only producer. More importantly, Du Pont was not the only one to work on CFC substitutes. There have been two main agents in the market for chemical products: the American Du Pont, and the British ICI. It is likely that even today, if one buys a chemical product – be it a paint, a spray, a glue or anything else – it will be an item produced in a plant owned by either of the giants. They can cooperate with each other, but first of all they see themselves as the most important market rivals. Economists look at their battlefield as a duopoly, i.e. a specific market structure whose dynamics depends on how the two main suppliers behave.

Both Du Pont and ICI understood that sooner or later freon would be substituted by better successors, and the winner would be the rival who can supply this new market easier or cheaper. Both firms have been involved in extensive research on freon substitutes. Du Pont was advanced in chemicals called *hydro-chloro-fluoro-carbons* (in short HCFC). ICI worked on chemicals called *hydro-fluoro-carbons* (in short HFC). HCFCs were closer to wide-scale industrial implementation, but HFCs were environmentally superior. The dilemma was thus whether to leave CFC-based technologies quickly by switching to HCFCs, or to postpone the improvement a bit, and switch to better technologies based on HFCs. The first solution would lead to adoption of technologies (economists call it "lock-in") that are inferior.

It is thus not striking that Du Pont and ICI occupied two opposite sides of a barricade. After the "conversion", Du Pont was supported by the American government and most environmental NGOs. At the same time ICI was supported by the European Commission and just a handful of environmental NGOs. Du Pont emerged as the winner. Yet the arguments were quite elaborate. They have been referred to by business administration specialists in academic textbooks, and their essence is captured in the table below.

Du Pont strategy		
CFC	**HCFC**	**HFC**
Keeping a high CFC price as a result of the immediate ban on its production, and supporting it by requiring CFC recovery and destruction	Stimulating irrevocable commitments of as many producers and users as possible, and thus creating *fait accompli*, as well as political support. Extend the period of using HCFCs	Discouraging expectations of early availability of HFC-based technologies
ICI strategy		
CFC	**HCFC**	**HFC**
Keeping a low CFC price as a result of postponing the ban on its production, and supporting it by requiring CFC recovery and reuse	Pressing the European Commission to ban HCFCs, making potential investors afraid of such a ban in the future, and thus reducing their number (which will make the ban perspective more credible). Shorten the period of using HCFCs	Vigorous investing in the new technology and hence creating political support for HFCs

Based on L. Gabel "Environmental management as a competitive strategy; the case of CFCs" pp. 328–346, in: H. Folmer *et al.* (eds.), *Principles of Environmental and Resource Economics*, Edward Elgar Publishing, Aldershot 1995.

As can be seen, Du Pont and ICI strategies were sophisticated and they were not confined to investing in chemicals that a given company was most advanced in. The strategies included shaping public opinion in such a way that governments are pressed to support a given technology. The ozone layer protection was referred to, but both firms cared above all about their market shares.

From the point of view of protecting the ozone layer, the dilemma was whether to limit the use of freon quickly but use a technology which is not very good, or to postpone the protection for some time and switch to a better technology.

Unfortunately, many environmental NGOs did not understand this dilemma. They seemed to be overwhelmed by a "victory" manifested by the main suspect, i.e. Du Pont, being converted.

The United Nations drafted a Convention on the ozone layer in Vienna in 1985. The convention was not very aggressive, but two years later – under pressure from the US government and applauded by environmental NGOs – a protocol was signed in Montreal to amend it substantially. The Montreal Protocol became an effective instrument to limit the production of CFCs immediately, just as Du Pont wanted. It is also true, however, that the Montreal Protocol was preceded by new alarming data about a drastic expansion of the ozone hole over Antarctica (Haas, 1992; Széll, 1993). The protocol set a schedule on limiting the global production of freon. The demand function in the market for freon resulted in a drastic price hike following a modest decline in supply. Consequently Du Pont recorded fantastic growth in profit. In response, the US government took a step that, in a market economy, can be taken only once every hundred years or so. Namely it introduced a so-called back tax (an *ex post* adjustment of tax rate) for Du Pont, which complained but had to admit that the unexpected growth in revenues resulted not so much from market developments as from a government decision. Therefore, if the government caused the exceptional revenues then the government can claim some of that money as well.

The Montreal protocol can be analysed as an instrument of a competitive strategy in a duopolistic market. Nevertheless it has become one of the most important global environmental agreements. It implemented certain unique principles that are not replicated elsewhere, but at the same time, it implemented some principles that established precedents and – for better or worse – are referred to when trying to solve other environmental problems. Seen from the perspective of more than two decades now, the protocol proved to be an effective tool in repairing the ozone "hole", but from the point of view of environmental protection it should be considered controversial. Promoted by Du Pont, HCFCs turned out to be greenhouse gases, i.e. substances that contribute to global warming. Hence the problem was not solved permanently; it was merely substituted with another one. Fortunately, the volume of HCFCs released to the atmosphere is small, so the damage done to the global climate in this way is small as well.

3.1.3 The Montreal Protocol

The Montreal Protocol was signed just two years after the Vienna Convention. It equipped the convention with "teeth" – i.e. specific provisions on how to limit the emission of CFCs – and defined who should do what. The protocol referred to 1986 as the base year and required (subject to some exceptions) that the use of freon is first frozen, and then decreased gradually. Developing countries were supposed to comply with the same regulations as those applicable to developed ones, except that the time frame for the entire programme was 10 years longer. The protocol proved effective at eliminating freon from economies. Occasionally American authorities report attempts to smuggle into the USA a container with freon that was illegally produced somewhere, but the scale of the problem is rather small. It is worthwhile to analyse principles of the Montreal Protocol. On the one hand, it proved to be an effective instrument to protect the ozone layer. On the other hand, it introduced some solutions replicated in other areas of international environmental cooperation where they are entirely ineffective.

First of all, the protocol divided the world into two distinct categories: developed countries and developing countries. According to the classification adopted by the United Nations Organisation, the latter are those that were considered unindustrialised in the 1980s, including Malaysia, South Korea, Mexico, and so on. Poland, Bulgaria, and other countries struggling with the heritage of the central planning system, belong to the former category. The protocol tries to maintain common sense, and grants developing countries certain privileges subject to the condition that their annual CFC consumption per capita is less than 300 g. By the way, consumption in Poland has always been much below this threshold, but it did not translate into any privileges.

The division of the world into two distinct categories – developed and developing – is sometimes defended on ethical grounds as a division into rich and poor. Of course, there is some rationality in this, but the dichotomy is too simplistic.

Introducing shrinking limits for freon use, the protocol allowed for some flexibility by making the limits tradable. As a matter of principle, countries (signatories) rather than firms are supposed to comply with the protocol. Trading is thus allowed between signatories. Governments have not used this option, since the limits turned out not to be very painful. In few cases – in the US, Canada, Mexico and Singapore – local systems of trading were introduced, having first allocated country-wide limits among domestic firms. It worked well, and, according to government reports, it helped to comply with requirements of the protocol (that were not very hard to begin with).

Another important provision of the protocol was establishing a financial mechanism. It let developed countries finance the so-called incremental cost of meeting the protocol requirements by developing countries. The concept of incremental cost was defined as the cost that a signatory is to bear in addition to what it would bear in the absence of the protocol. For instance, if an Indian entrepreneur estimates the cost of producing a CFC-free refrigerator at 400 dollars, then one needs to estimate the cost of producing the same refrigerator using the traditional (CFC-based) technology. If the cost turns out to be 350 dollars, then the incremental cost is 50 dollars, not the total cost of producing the commodity. In a special report prepared for protocol signatories, the incremental costs of all developing countries were estimated at 60 million dollars per annum. This is a relatively small amount that developed countries could easily agree to. An important outcome of the protocol was working out a methodology to estimate incremental costs.

Yet perhaps the most remarkable provision of the protocol involves trade sanctions. Namely, a signatory to the protocol cannot trade freon with a non-signatory. Since the USA is a party to the protocol, and every country would like either to buy freon from the Americans or to sell freon to them, then every country should be a party to the protocol. Thanks to trade sanctions, the process of ratification had no exceptions: all the United Nations members (197) are bound by the protocol.

Thanks to the Montreal Protocol, the depletion of the ozone layer has been stopped. Scientific investigations after the 2000 suggest that its size is not increasing. The concentration of CFCs in the atmosphere has stabilized and is even slowly declining. It is around 15% less than it used to be in the peak period in the first half of the 1990s, and it is below 2 PPB (*parts per billion*). It is expected that in the second half of the 21st century the ozone layer will be as it was in the first half of the 20th century.

The elimination of freon turned out to be easier and even cheaper than expected. The Montreal Protocol – and especially its flexible instruments such as emissions trading – should be credited for this. But first of all, freon turned out to be not nearly as irreplaceable as originally claimed. Many popular sprays are often advertised as "CFC-free". Initially it was believed that using a substitute gas as a propellant would require a higher cost. But, in some applications, freon was substituted by carbon dioxide or other inexpensive chemical compounds. It has been a surprise why producers failed to discover this earlier. Perhaps it illustrates the Porter hypothesis, which states that stricter environmental regulations do not impair industrial performance, but promote technological progress (Porter and van der Linde, 1995).

There were more surprises. For instance, freon used for cleaning microchips were substituted with an appropriate solution of soap, etc. In the short run, limiting freon use led to some economic disturbances. In the long run – say, five years and beyond – no economic hardships occurred.

Nowadays freon is a bit exotic, and is a relatively unimportant chemical. Its global use is negligible. Substitutes, i.e. HCFCs, pose a problem. They are indeed somewhat friendlier for the ozone layer, but they are powerful greenhouse gases. In the 1980s – when the Montreal Protocol was drafted and Du Pont tried aggressively to promote its product and downplay the research carried out by ICI – the problem was considered secondary. The protocol proved to be effective in getting rid of freon. From a longer term perspective, however, its environmental effectiveness has to be reconsidered.

The widely perceived success of the Montreal Protocol resulted in attempts to replicate its solutions in other areas of international environmental cooperation. It is thus worth looking at its provisions and analysing whether they can be replicated elsewhere.

Most importantly, the Montreal Protocol has the maximum possible number of signatories: all 197 United Nation members, even though parties have to accept certain commitments. This universal adoption was achieved thanks to trade sanctions to which potential non-signatories would be subject. As noted earlier, this was possible thanks to American participation and to America's crucial role in the CFC market. A similar provision cannot be replicated, as principles adopted by the World Trade Organization (WTO) exclude trade sanctions in environmental agreements.

Three articles have been commonly referred to when judging whether anything resembling trade sanctions can be tolerated: 3, 11, and 21. They read (GATT, 1947):

> *3. The contracting parties recognize that internal taxes and other internal charges, and laws, regulations and requirements affecting the internal sale, offering for sale, purchase, transportation, distribution or use of products, and internal quantitative regulations requiring the mixture, processing or use of products in specified amounts or proportions, should not be applied to imported or domestic products so as to afford protection to domestic production.* [...]
>
> *11. No prohibitions or restrictions other than duties, taxes or other charges, whether made effective through quotas, import or export licences or other measures, shall be instituted or maintained by any contracting party on the importation of any product of the territory of any other contracting party or on the exportation or sale for export of any product destined for the territory of any other contracting party.* [...]
>
> *20. Subject to the requirement that such measures are not applied in a manner which would constitute a means of arbitrary or unjustifiable discrimination between countries where the same conditions prevail, or a disguised restriction on international trade, nothing in this Agreement shall be construed to prevent the adoption or enforcement by any contracting party of measures:*
>
> [...] *necessary to protect human, animal or plant life or health;* [...]

The Montreal Protocol could have been legally challenged, but it was not, due to the minor costs its trade sanctions imply and due to the fact that incremental costs in developing countries are reimbursed. But if a measure is to imply a major burden, then the free trade argument is called forth to immediately.

All the widely publicized international court cases and legal proceedings leave no doubt that international trade is to be free from any environmental requirements. Despite that, repeatedly there are suggestions to ban, or to impose special tariffs on, products that violate environmental regulations adopted by the importing party. Voices making such suggestions become quieter when lawyers start citing WTO principles, but they eventually resurface. The Montreal Protocol has been an inexhaustible source of inspiration.

A good solution adopted by the protocol was emissions trading (strictly speaking, trading the freon limits allocated to signatories). The limits allocated turned out not to be very demanding, but the principle is correct. It is hence not surprising that emissions trading is added to many environmental agreements, including those that are aimed at carbon dioxide abatement. The principle is routinely called a "flexibility mechanism", since it relaxes the constraints which would bind every country individually, if the limits were non-tradable.

The principle that the rich finance the poor in order to let them meet the protocol requirements is remarkable. The world was divided into rich and poor. The former contribute to a fund and the latter draw from the fund. Financing is confined to so-called incremental cost, i.e. additional cost implied by adopting the protocol. This cost is not very high at the world scale. In 1990 it was estimated at 60 million dollars per year, and this estimate was a realistic one. The rich were supposed to collect this money. As a party to the protocol, Poland paid an annual fee of approximately 1 million dollars, which was not a big problem.

The principle that the rich compensate the poor is sound. The problem, however, is twofold. First, the poor are supposed to cooperate with the rich in order to achieve the common objective. Second, the rich should not hesitate to spend the money. Both requirements were met in the Montreal Protocol. The poor were supposed to limit the use of freons too, although at a somewhat slower pace. The rich paid what was necessary, as the sum was not excessive. Hence even the principle of a static division of the world into rich and poor did not prevent the protocol from being universally accepted by the former.

And yet this static division of the world into rich and poor is an unfortunate element of the Montreal Protocol. For two decades it has been responsible for the failure of international attempts to protect the global climate.

The dichotomous division of the world was adopted by the Montreal Protocol from the outset. Nevertheless, it was not supposed to be rock solid. It was conceivable to develop principles to draw the border line between rich and poor in a more rational manner, rather than perpetuating the status quo from the 1980s. The protocol did make the concept of a poor country slightly flexible by depriving certain privileges those whose annual consumption of freons was higher than 0.3 kg per head. However, it did not let leave the rich country category those whose annual consumption of freons was lower than 0.3 kg per head (i.e. those who were *de facto* poorly developed). As a result, Poland – even though it had annual freon consumption lower than 0.3 kg per head – was considered a rich country with all the practical consequences of this status.

In the spring of 1990 there was a meeting of parties of the Montreal Protocol in Geneva where I represented the Polish government. I tabled a motion that countries which had just abandoned the central planning system – so-called transition economies – were given special status under the protocol (neither "rich" nor "poor"). A representative of the Soviet Union who seemed to understand the problem quietly supported this motion. But the Soviet Union was in the middle of a civil war, and its representative was apparently instructed to keep a low profile and not to cause a disturbance. Representatives of other transition economies did not seem to understand what the issue was. Consequently my motion did not gain a broad support, though a representative of India took note. In the plenary meeting, he angrily demanded that the Polish party withdraw it from the minutes. Representatives of European Union countries applied some pressure, arguing that Poland was to receive a great deal of assistance money anyway, so its potential financial contribution would not be onerous.

The motion was not withdrawn. The Secretariat of the protocol repeatedly asked the Polish government about specific modalities to be implemented in order to develop the motion. However, at that time, Poland was applying for membership in the Organization for Economic Cooperation and Development (OECD), considered a "wealthy-country" club. Thus, suggesting that this "wealth" is relative would have been awkward. The issue died naturally, and amendments to the protocol adopted in London (1990) and Copenhagen (1992) did not create a separate category for transition economies.

The Montreal Protocol set a strict division of the world into "rich" and "poor". The division was not a problem as long as the "rich" did not have to pay too much for the "poor". Protection of the ozone layer is not a typical environmental concern cost-wise. Substituting freon with better substances (especially following the scenario inspired by Du Pont) turned out to be rather cheap. As a result, the cost of the protocol is fairly low, so the "rich" do not have to pay much in order to finance incremental costs incurred by the "poor". Nobody protests against excessive expenditures. Other environmental programmes are different. Their costs are higher. The "rich" protest when their money flows to the "poor" who – as it turns out – can be wealthier than the donors. By referring to the Montreal Protocol, the "poor" insist that this is how environmental agreements should be arranged. It is difficult to depart from this principle.

3.2 Climate

Geological records illustrate drastic changes that the Earth climate was subject to long before the presence of man and the human economy. There were both glaciations and tropical temperatures. What is unique now is the pace at which changes occur. If a temperature or precipitation changes slowly, then organisms that do not tolerate new circumstances can migrate (either southward or northward). But if circumstances change quickly, then there is no time for migration, and certain species may simply become extinct. The climate changes observed in the 19th and 20th century are fast and are therefore a matter of concern to many scientists.

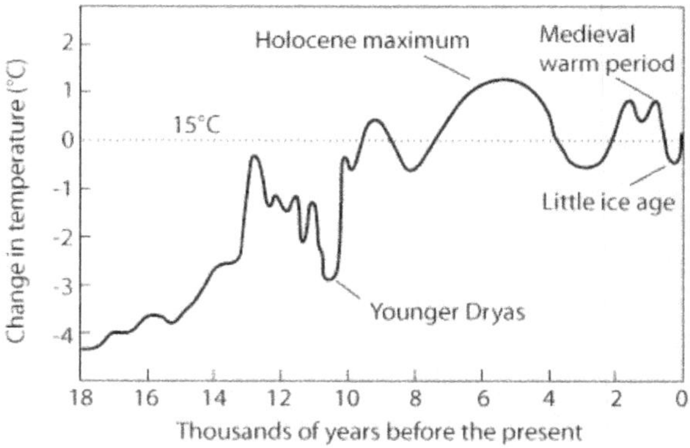

Source: Marshak (2007)

There is serious discussion about whether these changes have been caused by human activities originated from somewhere else. There is also a serious discussion about the future of such changes. Scientific opinions are not unanimous, as the issue is an extremely complicated one. There are many climatology experts, but knowledge of how the chemical composition of the atmosphere is linked to the Earth's temperature is far from common. It is even worse when scientists try to numerically solve gigantic systems of partial differential equations that simulate the Earth's climate. There are only few computer models in the world which accomplish this up to adequate precision standards.

There is no certainty, but most specialists share the opinion that the expected growth of atmospheric concentration of carbon dioxide will imply the growth of Earth's surface temperature by a couple of degrees Celsius by the end of the century. The most spectacular effect of this process would be melting the arctic ice and rising sea levels. But there are other anticipated consequences as well, one of which would be a change in the pattern of extreme weather conditions, such as droughts and floods. Another one would be a changed frequency of hurricanes and heat waves. Changing the circulation of the Gulfstream (which now cools down North America and warms Europe) would be a total disaster. It is simply hard to predict all the consequences of such a development.

The geographical distribution of expected losses is not uniform. What is certain is that some small island states in the Indian and Pacific Oceans would be submerged under water. Agriculture in Africa would be harmed. Some areas – such as parts of Siberia and North Canada – would perhaps record benefits. There is no certainty, but it is assumed that worldwide losses would dominate over gains.

Whenever and wherever there is a flood, a drought, or a heat wave resulting in a higher death toll (especially among urban populations), journalists suggest that the problem has been caused by climate change. This is a simplistic argument since no short-term observations can tell anything about climate change. At most, one can complain about the weather. Making conclusions about climate change requires observations that last many generations and takes into consideration long-term average values in order to exclude short-term fluctuations.

3.2.1 The greenhouse effect as a public good

Scientists can draw from long-term observations of atmospheric carbon dioxide concentration and average temperatures of the Earth surface. The former is measured in Parts Per Million (PPM). It ranged between 280 in the first half of the 19th century to around 400 in 2014. Based on various scientific (e.g. geological) records, it is estimated that over the last several thousand years it was between 200 and 280. This data is confronted with records of average Earth surface temperature measurements. In the 1950s it was 0.5°C higher than over the previous thousand years. The most spectacular growth took place in the second half of the 20th century. It is almost 1°C, if one refers to 10-year average values (which eliminate annual random fluctuations).

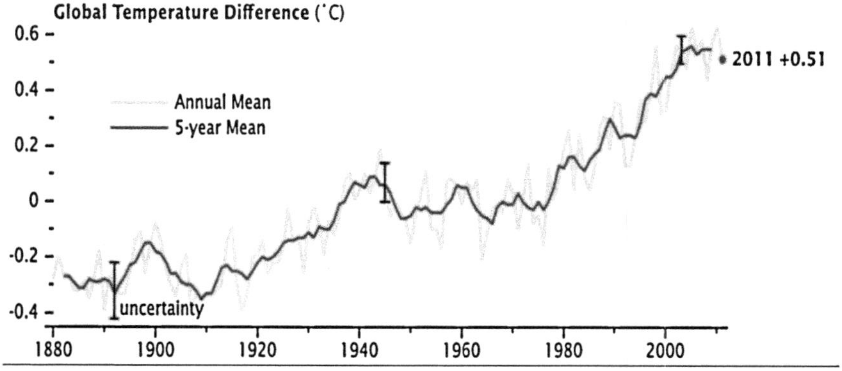

Source: NASA (2012)

The direction of the causal relationship (and even the mere existence of such a relationship) is debatable. There are analyses which suggest that changing concentrations result from, rather than cause, changing temperatures. Yet the reverse opinion is prevalent. A vast majority of specialists claim that the growing concentration of carbon dioxide causes rising temperatures. A theory which explains this was formulated by Svante Arrhenius, a Swedish chemist, and it has been known since the end of the 19th century. It is based on the fact that carbon dioxide is more transparent with regard to ultraviolet radiation than it is to infrared radiation. At the same time, incoming solar radiation includes a large share of ultraviolet light, and the radiated back energy includes a large share of infrared light. Carbon dioxide acts thus like a trap for energy: it lets it reach the Earth surface but does not let it return into outer space. For the planet, in order to achieve thermic equilibrium (so that the amount of incoming and outgoing energy is the same), the temperature – which determines the amount of radiated energy – must increase.

This is the so-called "greenhouse effect", which gardeners have utilized for many years. They cover their flower or vegetable beds with something in order to increase the temperature as a result of increasing local carbon dioxide concentration. Thanks to such tricks, plant cultivation becomes faster or more efficient. The Earth's atmosphere acts as a huge greenhouse. Thanks to the trace presence of carbon dioxide it does not let the planet surface cool down immediately after the sun sets but keeps the temperature more stable, thus allowing for life as we know it. What scientists are concerned with is the unusually fast pace of the increase in carbon dioxide concentration.

Carbon dioxide is a chemical substance that exists in natural systems. It is emitted in the process of breathing, so some quantity must circulate in the atmosphere anyway. Many millions of years ago (when the climate was much warmer) its atmospheric concentration was much higher. Carbon dioxide circulated in the biosphere and allowed magnificent forests to function. Because of reasons which are not quite clear, an ecological catastrophe occurred that resulted in the death of these forests. They were petrified and covered under the Earth surface. Just recently they were discovered as fossil (carbon or hydrocarbon) fuels. By burning them the man, in just several centuries, brings the carbon dioxide back to the atmosphere that was eliminated from it in substantially longer geological epochs.

Burning fossil fuels is not the only source of carbon dioxide. Deforestation acts in a similar way. As long as a forest lives, carbon compounds are stored in trees. If the forested area shrinks – and this is what is happening in the world today, especially outside of Europe – this inventory is threatened. As long as wood keeps its integrity, e.g. in the form of products of wood industry, the inventory is maintained. But as soon as disintegration occurs, and any product will disintegrate sometime, the inventory goes back to the atmosphere as carbon dioxide.

Several other gases behave in the atmosphere like carbon dioxide. They are called "greenhouse" gases (GHG). Methane is the most important item on this list. It is emitted in the course of cattle husbandry and rice cultivation. Its emission is much lower than the emission of carbon dioxide, but its impact on climate (so-called "forcing" potential) is much stronger. Thus its impact cannot be considered negligible. GHG include some of the freon substitutes supported by the Montreal Protocol.

Global climate is a textbook example of a public good, since many users can benefit from it at the same time and nobody can be excluded from its benefits. Everybody will benefit by protecting the climate. But (like any GHG) carbon dioxide is perfectly mixed in the atmosphere. Its origin does not play any role. Therefore if a region decreases emissions of carbon dioxide, then it protects the climate for everybody, not only for itself. And *vice versa*, if a region continues emissions, then it causes damage for everybody, not only for itself. Therefore if a region abates but the global emission increases, then it pays the cost of its own abatement, but in addition it suffers damage that it was not necessarily responsible for. There is a sort of joint responsibility in the economics of public goods: one bears consequences of someone else's actions or inactions.

Failure to understand the nature of a public good results frequently in a wrong argument applied by climate protection advocates. For instance, they argue that abatement cost is lower than the loss resulting from non-abatement. This may be a true statement but only at the global level. If a region abates unilaterally then it bears the cost of abatement. Yet losses depend not on its local emission, but on what other regions do (or fail to do). That is why it is absolutely crucial that climate protection is undertaken globally rather than unilaterally.

Climate protection has been formally recognized as a global problem. A United Nations Framework Convention on Climate Change (UNFCCC) was publicized in Rio de Janeiro in 1992, which is also known as the "Rio Convention". This is not quite correct, as the text – after a long negotiation process – was opened for signature earlier at the UN in New York. Hence it should have been called the "New York Convention".

The convention proved totally ineffective, because in the two decades since its adoption, the global emission of carbon dioxide has increased by 40% (from around 25 billion tonnes to around 35 billion tonnes). These statistics are approximate, because (unlike LRTAP) the convention did not establish a reporting obligation. Only a narrow group of countries – so-called Annex I – made up of 40 economies considered "rich" (this resembles the Montreal protocol) is supposed to report GHG emissions. At the turn of 1980s and 1990s they were responsible for a high percentage of global emissions. Now they emit much below half of the total. In 2006 China surpassed the USA, and its "advantage" has been growing fast. But Chinese emissions – like those from all the other (more than 150) non-Annex I countries – are only estimates, because official statistical sources are unavailable. There is also a discrepancy between carbon dioxide emissions and all GHG emissions calculated in CO_2 equivalents. Finally, there is a discrepancy between emissions from industrial processes (mostly fossil fuel combustion) and total emissions taking into account land use changes (mostly deforestation).

The present concentration of carbon dioxide results not only from current emissions, but rather from cumulative emissions. Cumulative emissions – calculated typically from the middle of the 19th century – originate mainly from countries which became industrialized earlier. As a result, the US, European Union countries, and China are responsible for 19%, 17%, and 12%, respectively. However, given the projected drastic further emissions growth from China, and a significant decline from the USA, in the 2020s the shares are likely to converge, perhaps to a common level of 16%.

On a per capita basis, annual global emissions of GHG is around 6 tonnes. Including estimated emissions from land use changes increases this number to 7 tonnes. It is roughly equal to what is emitted in China now. In Poland it is slightly below 10 tonnes, and the average in European countries is somewhat higher. Oil extracting countries from the Persian Gulf are on the very top of this ranking (more than 30 tonnes). Out of 86 countries which emit more than the average, a vast majority do not belong to the Annex I, and consequently they do not have any reporting obligations. Neither do they have any commitments to limit their GHG emission within the UNFCCC.

Enthusiasts of unilateral reductions of carbon dioxide emissions think that they protect the global climate. But they do not. The more a region abates unilaterally, the higher the emission can be elsewhere. European enthusiasts of unilateral abatement complain that whatever they do, global emissions have grown. In fact, this outcome is not despite, but rather a result of their actions. So-called *carbon leakage* is responsible for it. An abating region does not necessarily free the planet from excessive emission, as the emission may simply move elsewhere.

Empirical research carried out in Great Britain – which would probably deliver similar results in many other countries – demonstrated that direct carbon dioxide emissions have declined over the last two decades, but only because local production was substituted with imports. Nevertheless enthusiasts of unilateral abatement question whether such calculations can be explained by *carbon leakage*. They suggest that perhaps the British – independently of any climate actions – changed their preferences in such a way that they buy more from regions which do not impose any constraints on carbon dioxide emissions. From the purely logical point of view, such an explanation cannot be rejected. However, economists identified several mechanisms that may convincingly explain why unilateral abatement leads to increased emission elsewhere (see section 3.2.4).

Supported by some Europeans, the concept of unilateral abatement has diverse justifications. One which is closest to a rational argument is the expectation that technological progress triggered by such an activity will lower the cost of carbon-free production to such an extent that countries which do not have any constraints on emission will abandon old production techniques and switch to new ones. Technological progress does take place indeed, but its pace is much slower than expected, so that old "dirty" methods are still cheaper. As an alternative, an ethical argument can be used: the rich accumulated a debt to the poor (by spoiling the climate) starting in the 19th century, so they should pay it back now. However, estimates of the cumulative share in carbon dioxide concentration in the atmosphere (just quoted above) demonstrate that the role of the (formerly) poor is quite sizeable. Hence climate protection requires that the poor accept responsibility, too. But when all the arguments that refer to certain facts fail, a *leadership* argument is evoked. "Europe should serve as an example for the rest of the world". Inspiration provided by a good example is an excellent method of raising children. But it fails as a method to carry out complicated international negotiations.

The literature on the economics of climate change is far from rigorous. In the 1990s it included papers on alleged rationality of hurting yourself as a means to increase your credibility in the eyes of your adversaries. While it certainly makes sense in some circumstances, it is not relevant for climate negotiations, which suffer from unfortunate UNFCCC provisions rather than the poor reputation of some negotiators. Today, the literature includes much more relevant contributions, but it is also abundant in professional papers that address sophisticated theoretical problems with little reference to the most salient policy questions. One of the best known experts in the field, Richard Tol, writes about "the hubris that pervades climate research and policy" (2014). When looking who is guilty for the lack of climate protection, it is easy to blame the Americans or the Chinese. However, one needs to raise an inconvenient question, whether doing what seems to lead "in a good direction" is indeed beneficial. It turns out that this logic may not work. What is marketed as "climate policies" may lead to deforestation, biodiversity loss and other calamities.

Recent decades mark a deterioration in the climate change predicament. Even in regions, such as Europe, where the problem is high on the political agenda, economies are failing to make a serious effort to abate carbon dioxide emissions. An assessment made by the European Environment Agency demonstrates that carbon dioxide emissions released by the steel and cement sectors are falling much slower than production (EEA, 2013). In other words, pollution per unit of output has increased after 2008. The message gets even worse. As consumption of cement and steel products is likely to grow in low and medium income countries (it has stabilized in the most developed economies only), lower domestic production implies higher production elsewhere, i.e. a mere relocation of emissions.

3.2.2 The Intergovernmental Panel on Climate Change

The Intergovernmental Panel on Climate Change (IPCC) was established in 1988. Almost all countries of the United Nations system (195) participate in the work of this body, which was instrumental in drafting the UNFCCC adopted in

1992. It has a generous budget, although its amount is a matter of controversy and endless speculation. The official number for the 2008 revenue was 6 million Swiss francs, i.e. somewhat less than $6 million (IPCC, 2009). However, the total number is difficult to ascertain, since specific activities are separately funded through dedicated grants. The cumulative sum for the initial two decades of the operations is over $100 million. But even this number may be misleading, as financing of some projects is carried out by "rich" country governments directly. While the total funding is thus not clear, it is obviously much below what can be found in the blogosphere.

The IPCC is a powerful body and its success made environmental campaigners active in other (not climate-related) areas envious. Scientists working in air pollution, water abatement, biodiversity preservation etc. would be happy to participate in programmes as abundantly funded as those organized by the IPCC.

According to its mission, the IPCC provides scientific information to governments. The best known forms of its activities are *Assessment Reports* published every several years. The most recent one (IPCC, 2013 and 2014a-b) was prepared in 2012–2014. The previous one (IPCC, 2007a-c) was published seven years earlier. These reports consist of three volumes, each summarizing the outcome from one working group. The Working Group I deals with *Physical Science Basis*, II – with *Impacts, Adaptation and Vulnerability*, and III – with *Mitigation of Climate Change*. The entire report is an ambitious endeavour, with more than 800 authors involved in its preparation.

Report writing is an example of an activity which certainly requires funds, though its budget is difficult to establish. With a large number of authors from all over the world, and their multiple meetings, it is an expensive endeavour. Authors coming from developing countries are funded by dedicated grants offered by the "rich". Other authors are funded by their respective governments directly, and it would require enormous work to compile information on the total cost.

Authors invited to draft the reports are instructed to examine the most recent peer-reviewed literature in order to inform governments about what the scientific community has to offer with respect to climate change. There is a diversity of scientific opinion on policy options. However, even in the Working Group I almost every claim can be found in peer-reviewed literature. In order to communicate uncertainty, the IPCC (2010) developed special terminology which the reports are to apply. First of all, authors are supposed to make a distinction between evidence (which can be "limited", "medium" or "robust") and agreement, related to the interpretation of the evidence (which can be "low", “medium” or “high”). In addition, authors are supposed to use words such as “likely” or “unlikely” in a very precise way. The following table lists expressions to be used.

Likelihood of the outcome [%]	Expression to be used
99–100	virtually certain
90–100	very likely
66–100	likely
33–66	about as likely as not
0–33	unlikely
0–10	very unlikely
0–1	exceptionally unlikely

A serious effort was thus put into making the reports as precise and practical as possible. Nevertheless, contributions from Working Groups II and III are doomed to be less universally accepted than those from Working Group I. The former may depend on the choice of authors to a larger extent than the latter.

The process of appointing authors has not been very clear. I accepted an invitation to contribute to Working Group III, and I was honoured to cooperate with colleagues to write Chapter 15 of the report (entitled *National and Sub-national Policies and Institutions*). Based on my experience, I consider the report a good review of the scientific literature. Nevertheless, given the fact that, in policy papers published in good international journals, almost any claim can be found, report authors should make a conscious selection of the material, and inform policy makers according to the best professional knowledge they represent. In principle the report writing process was arranged in a way which allows for adequate interpretation of the literature. Subsequent drafts of the report were reviewed by hundreds of scientists and policy makers. Authors were supposed to respond to, and possibly incorporate, myriads of comments. Chapter 15 received more than a dozen comments per page. Some of them contradicted each other, but some included pertinent thoughts and helped in the revision of the text.

However, given the number and inconclusiveness of these interventions, it was obvious that the IPCC Secretariat would be instrumental in preparing the final version of the report and its message. According to many environmental economists, Chapter 15 should reinforce the message about the public good nature of climate protection. Consequently it should stress that national and sub-national mitigation policies cannot substitute for a global mitigation policy. Instead, Chapter 15 includes a great number of statements about how such sub-global policies relate to the (non-existing) global one. Having received a long list of detailed facts, readers may be confused as to the main message.

Despite these observations, IPCC reports provide a sound basis for developing knowledge on climate change. Unfortunately, they are considered too long to be read by decision makers. Each chapter is supplemented with an "executive summary". These summaries are of course much less informative. But the main problem is in even shorter "summaries for policy makers" (SPM). They are to be approved by all (!) signatories of the UNFCCC. As a result, SPMs are washed out of whatever is meaningful, but may be found cumbersome by a single government. The SPM contains statements which everybody can agree to, and is consequently of little practical value (Stavins, 2014).

IPCC reports are widely read and commented on. As in the past, some readers point to a few factual errors, but the most common reaction is a complaint that – despite the high cost of inaction – the world continues its inefficient trajectory. While this is true, it misses the point about the main cause of the failure, i.e. the lack of a global agreement.

3.2.3 The Berlin Mandate

Soon after the convention came into force in 1994, at the first official meeting of parties in Berlin in 1995, the signatories agreed on a number of important amendments. The most remarkable of these – known as the Berlin Mandate – states that only Annex I countries will take binding commitments. The wording may not be quite clear, but non-Annex I countries insist that this has been the intention of the amendment:

The Parties should protect the climate system for the benefit of present and future generations of humankind, on the basis of equity and in accordance with their common but differentiated responsibilities *and respective capabilities. Accordingly, the developed country Parties should take the lead in combating climate change and the adverse effects thereof.*

The principle of "common but differentiated responsibilities" (CBDR) was adopted as the cornerstone of subsequent activities under the UNFCCC.

The signatories meet annually at plenary meetings called Conferences of Parties (COP). Between 1995 and 2014 there were 20 such events. The list is as follows:

- COP-1 Berlin (1995)
- COP-2 Geneva (1996)
- COP-3 Kyoto (1997)
- COP-4 Buenos Aires (1998)

- COP-5 Bonn (1999)
- COP-6 Hague (2000) / Bonn (2001)
- COP-7 Marrakesh (2001)
- COP-8 New Delhi (2002)
- COP-9 Milan (2003)
- COP-10 Buenos Aires (2004)
- COP-11 / MOP-1 Montreal (2005)
- COP-12 / MOP-2 Nairobi (2006)
- COP-13 / MOP-3 (2007) Bali
- COP-14 / MOP-4 (2008) Poznan
- COP-15 / MOP-5 (2009) Copenhagen
- COP-16 / MOP-6 (2010) Cancun
- COP-17 / MOP-7 (2011) Durban
- COP-18 / MOP-8 (2012) Doha
- COP-19 / MOP-9 (2013) Warsaw
- COP-20 / MOP-10 (2014) Lima

The third Conference of Parties adopted a famous decision on limiting GHG emissions. The Kyoto Protocol came into force in 2005, and subsequent COPs serve as "Meetings of Parties to the Protocol" (MOPs) too. The Kyoto Protocol embodied the principle of CBDR. Only Annex I countries took binding commitments to limit GHG emissions. The commitments were not spectacular (on average, a decrease of 5.2% from the baseline of 1988–1990 by the year 2012), but they confirmed the unfortunate principle.

Neither the convention nor its subsequent decisions succeeded in stabilizing carbon dioxide emissions. On the contrary, they have continuously grown, and the pace of this process has accelerated. As non-Annex I countries do not report their carbon dioxide emissions, it is difficult to assess global emissions. Nevertheless, the European Bank of Reconstruction and Development (EBRD, 2011) collected data to conclude that the annual rate of growth in carbon dioxide emission prior to the UNFCCC was 0.6%, then it increased to 1.2%, but after the Kyoto Protocol it rose to 2.6%. Of course, there could have been dozens of additional reasons for this trend, but it clearly indicates that non-Annex I countries' emissions are growing at an ever increasing pace, which makes climate protection targets unachievable. More recent data (IPCC, 2014b) show that the average annual growth rate of all GHG emissions was 1.3% in 1970–2000, and 2.2% in 2000–2010.

In 1989 it was projected that excluding developing countries from global action would result in a 40% increase in greenhouse gas emission by 2050 (Lashof

& Tirpak, 1989). This, however, happened much sooner. Global carbon dioxide emission grew by 40% between 1992 and 2012 already.

The Kyoto Protocol was agreed upon late in the night, just hours before the closing ceremony of COP-3. It was signed by all the Annex I countries. However, the US government signed it knowing that it would not be ratified by the Senate. Indeed, when ratification was put to a vote a couple of years later, out of 100 senators 95 were present, and the vote was unanimous: 95 against, and 0 for. Interestingly, Al Gore voted against the Kyoto Protocol, even though he forgot about this fact later on.

In the American parlance, the Kyoto Protocol is a "bipartisan issue", i.e. an issue that both parties (the Democrats and the Republicans) agree upon. In this case, they are both against. One can look for sinister reasons to obstruct, but another reason is that simply the Kyoto Protocol – as an agreement based on the Berlin Mandate – cannot be effective, and the sooner parties abandon this logic the better for climate protection.

Despite the American position, the European Commission struggled to defend the Kyoto Protocol. In order to become effective, the protocol was to be ratified by 55 parties, among which Annex I signatories should account for at least 55% of emissions from the Annex I. The number of ratifications proved to be easy. It exceeded the threshold by the early 2000s. However, once the USA announced its no-ratification decision, it was clear that the 55% threshold could not be achieved unless Russia ratifies the protocol.

The Russian government was aware of its negotiating power and did not want to take a decision before all the problems that could have been politically debated at the European forum were solved to its satisfaction. As a result, between 2001 and 2004 there were several contradicting press releases about Russia's willingness to ratify the protocol. Once all the problems were sorted out, the protocol entered into force in 2005, following Russia's ratification in the late 2004.

The European Commission saw this as a step in the right direction. As statistics show, it did not move us in the right direction, though some analysts have thought it did. The Commission immediately launched an emissions trading system (ETS) in order to make the Kyoto commitments easier for its members. The appropriate directive (2003/87/EC) was adopted even earlier, as the European Commission anticipated Russian ratification. The preamble says:

> *This Directive aims to contribute to fulfilling the commitments of the European Community and its Member States more effectively, through an efficient European market in greenhouse gas emission allowances, with the least possible diminution of economic development and employment.*

Based on the preamble, to some member states it was clear that meeting the Kyoto Protocol requirements is a reason to be exempt from some of the directive's obligations. But it proved not to be the case. Environmental economists have known that emissions trading is an excellent mechanism to minimize the cost of achieving an exogenously (and *a priori*) adopted emission target. As the instrument has been in wide use outside Europe since the 1970s, it was thoroughly researched and included in environmental economics textbooks. In particular, the texts elaborated on typical mistakes that a successful implementation of the instrument should avoid. Unfortunately, the ETS committed these mistakes, like changing allocation rules when the system is in operation, or adding additional instruments that undermine the trading. For instance, promoting certain technologies which decrease emissions is counterproductive, as it simply allows other emission sources to use up the space just freed. The European Commission is becoming aware of such problems and it is trying to correct its policies by establishing new rules, introducing new regulations, further manipulating with carbon dioxide allocation and thus creating new problems. The ETS suffered also from a number of technical problems. For instance, VAT refunds were fraudulent, before VAT harmonization was enacted.

In some countries – like in Poland in the 1990s – where expert and public opinion was positive about emissions trading, the ETS became synonymous with an unfair bureaucratic instrument which also fails environmentally. Environmental activists blame the ETS for not delivering expected climate protection measures. This is not a justified indictment, since the role of the ETS is not lowering the emission, but rather achieving cost-effectively whatever emission target was politically adopted.

The Kyoto Protocol has not helped climate protection, and it could not do so as an agreement based on the Berlin Mandate. Many years after its drafting (and entering into force) the global emission of carbon dioxide is much higher than it was. But the worst outcome is the waste of precious time. Creating an unfortunate impression that something can be achieved without binding commitments taken by non-Annex I countries resulted in a much more difficult negotiating predicament than it was in the 1990s. This has been aggravated by the fact that some constituencies found a number of climate-motivated regulations (ineffective from the climate protection point of view) financially attractive, and they try to prolong the existing arrangements. In particular, climate negotiations are influenced by various lobbies who are either for or against carbon taxes (Grazia *et al.*, 2014).

Even though the Kyoto Protocol petrified the CBDR and hence failed to limit global emissions, it did include a few good principles. Most importantly, it allowed for some flexibility in abatement thanks to emissions trading. Three articles – 6, 12, and 17 – include provisions on emissions trading. Article 6 simply allows emissions trading. Article 12 – advertised as an outreach to non-Annex I countries – allows Annex I countries to claim emission reduction credits created through abatement in a non-Annex I country; so-called Clean Development Mechanism, CDM. This sounds excellent, except that it does not contribute to lowering global emissions, since non-Annex I countries do not have binding commitments (see 3.2.4). Therefore they can expand their emissions of GHG, despite the fact that each CDM project indeed lowers a hypothetical emission that could have occurred in the absence of this project. The so-called Joint Implementation (JI), i.e. claiming an emission reduction credit by an Annex-I country created through abatement in another Annex I country, is referred to in Article 17. In addition, the Kyoto Protocol somewhat weakened the dichotomous division of the world into "rich" and "poor" (inherited from the Montreal Protocol), by allowing flexible baselines. In principle, all Annex I countries were supposed to take 1990 as the base year for calculating reductions. Included in the Annex I group, economies in transition were allowed to choose alternative baselines. Poland took advantage of this provision by choosing 1988.

The predicament is grave, but not entirely hopeless. In order to be environmentally effective, an agreement that is called for should imply binding commitments by non-Annex I countries. In order to be economically sound, the agreement must allow low cost abatement options to be taken before high cost options. In order to be fair, the agreement must make the "rich" pay for the "poor". An emissions trading scheme is an excellent instrument capable of delivering all three objectives. It is, however, unlikely that the scheme developed under the 2003/87/EC Directive will become a pattern for a global mechanism. Perhaps some other schemes – such as those adopted in some regions of the USA – may prove to be more appropriate.

The logic that seems to be adequate to help climate protection is based on recognition of the facts that every country needs to face the same incentive to abate, and that the "rich" should pay for the "poor". Assuming – for simplicity – that one looks for a fair allocation of the present global sum of carbon dioxide emissions, let us put $x_1+\ldots+x_k = 42$ billion tonnes. The more difficult task is to suggest an allocation principle. One sometimes referred to is based on population. Assuming further that the number of people in the world is approximately 7 billion, this gives an allowance of 6 tonnes per person. Then the allocation principle can be simply $x_i=6L_i$, where L_i is the population of the *i*th country. This would leave the European Union and the United States with allocations much below their current emissions; China – more or less at the level of current emissions; and other developing countries – much above. The tradability of permits would then imply a flow of wealth in a direction that is consistent with popular equity convictions.

There are conceivable alternative allocation principles. Some analysts advocate for distributing carbon dioxide emissions in proportion to GDP. This would leave the European Union and the United States with a much higher allocation than under the previous scheme. However, it drastically violates the CBDR philosophy, since those with higher GDP should not assume lower but rather higher abatement commitments. Arguments used at COP meetings prove that the notion of CBDR is present in all the discussions, but parties do not want to acknowledge that fairness implied by it eliminates effectiveness, which requires that also non-Annex I countries take binding commitments.

The adherence to the Berlin Mandate is the main reason for failure of global climate protection. Allowing non-Annex I countries to increase their emissions freely implies that the global emission would grow even if Annex I countries reduced their emission to zero. This further suggests that Annex I countries should commit to have negative emissions. It does not make economic sense, but this is how the concept of carbon capture and storage (CCS) was invented.

CCS is an idea to withdraw CO_2 from the atmosphere by e.g. pumping it under the earth surface. It may have some justification in hydrocarbon extraction, as something needs to be pumped in order to use up the deposit more efficiently. It makes no sense in coal mining. Despite that, the European Commission introduced a directive – called CCS-ready Directive (2009/31/EC) – which requires new coal-fired plants to be "CCS-ready". This means that the plants need to have infrastructure to capture CO_2 and pump it somewhere.

CCS is ineffective as a way of significantly reducing carbon dioxide emissions. Contrary to what some people believe, emissions cannot be transported and pumped into places from which fossil fuels are extracted. High-school chemistry demonstrates that combustion transforms each atom of C into CO_2, i.e. a molecule of the weight of 12 into a molecule of the weight of 44. In other words, combustion leads to an increase in mass by 3.67 times. Carbon dioxide is a gas whose volume is thousands of times larger than the volume of coal. It can be liquefied when the pressure of 200 bars is applied, but still its weight is almost four times the weight of the coal combusted. Burning oil or natural gas instead of coal changes this ratio somewhat, but the bottom line is the same: the mass of carbon dioxide is much larger than that of the original fuel. Thus arguments that "carbon dioxide can be pumped through a pipe parallel to the original one" is based on a misperception. Likewise, arguments raised by proponents of the CCS-ready Directive who claimed that "carbon dioxide can be transported back by the same trains which bring the fuel" are wrong. The CCS idea requires a transport infrastructure four times as large as the fuel provision. Members of the European Parliament who voted for the directive based their beliefs on completely wrong convictions. While it is likely that many of them simply did not understand the problem, there must have been some who voted for this idea just to claim that "negative" emissions are possible. Indeed they are, but this is an extremely costly and ineffective way to try to reconcile the Berlin Mandate with climate protection.

After a couple years of allocating large public funds to experiment with CCS and having learnt that private investors would not choose this technology (even with subsidies), governments stopped their funding. Some environmental NGOs who believed that CCS would help climate protection seem to be upset, but the technology is so bad that it could have been contemplated only as a desperate defence of the Berlin Mandate (Braend, 2014).

Geoengineering is another idea on how one may protect the climate if someone else keeps spoiling it. The idea was born many years ago (NAS, 1992), but the Berlin Mandate provided incentives to explore its many variants, some of which are more than controversial. For instance, an idea has been contemplated to contaminate the atmosphere with some aerosols in order to decrease the flow of solar energy reaching the Earth surface. The IPCC (2007c) concluded that these are largely speculative options. While the IPCC (2014b) did not endorse geoengineering as a solution to the climate protection problem, the mere fact that it resurfaces over and over again is an unfortunate effect of failure to acknowledge that limiting global carbon dioxide emissions is the only sound way to address climate change.

3.2.4 Carbon leakage

The UNFCCC's failure to protect the climate is sometimes attributed to the fact that a number of developing countries embarked on economic growth trajectories in an attempt to replicate what "rich" countries did in the 19th and 20th centuries. Of course, global development trends contribute to the fast growing emission of GHG, but the UNFCCC has not implemented measures to control climate change effectively.

One effect of the Berlin Mandate is some Annex I countries' willingness to undertake unilateral abatement actions in the hope that their moves will lead to lower global emissions. This would have been the case if emissions from a non-abating region were independent from what the abating region does. But in general this is not the case. In what follows, carbon leakage (CL) is defined as an emission in one geographical area resulting from a decrease in emissions elsewhere, everything else being constant, including policies elsewhere. It may happen that these policies elsewhere may change as a result of inspiration caused by one region's unilateral abatement policy, but this is not assumed here. Let N denote the region where the carbon emission "leaks to", even though N may undertake some abatement, and A denote the region which contemplates undertaking a more ambitious abatement programme. Formally CL can be defined as the difference between the expected emission in region N if there is an additional abatement programme in region A, and emission in region N provided standard ("business-as-usual") policies in region A:

$$CL(\Delta R) = (f_N(GDP_N,P_N,GDP_A(R_0+\Delta R))-f_N(GDP_N,P_N,GDP_A(R_0))) \,/\, \Delta R$$

where R_0 is the baseline (standard) reduction target adopted in A, ΔR is an additional reduction target contemplated in A, P_N identifies an abatement policy adopted in N, f_N is an emission function for N, GDP_A is a function of a reduction target adopted in A. This definition of carbon leakage allows for any sign and any level of the indicator. In particular, $CL(\Delta R)$ can be either positive or negative.

A negative value of CL (emission reduction in N corresponding to an increased carbon abatement target adopted in A) may arise as a result of complex processes that take place in the world economy following an increased abatement effort in one region. For example, this occurs if the transition of A to a low-emission economy induces a strong technological progress which is then adopted in N. In such a case, increased global output could be accompanied by less pollution globally. A positive value of CL (emission increase in N corresponding to an increased abatement target adopted in A) may result from moving production to where it is not constrained by environmental standards.

In this definition it is crucial that only R and GDP_A change. However, it is very difficult to comply with this assumption since all variables are linked to each other. GDP and emission in each region may depend on reduction targets in several ways. First, increasing reduction targets is typically associated with a slowdown in GDP growth. Second, increasing reduction targets in A is expected to decrease global prices of fossil fuels; this is likely to increase the demand for fossil fuels in N and to increase emissions in N (this effect is not observed in regions with binding commitments). Third, technological progress and economies of scale in low carbon technologies are likely to drive their costs down (both in A and N). Fourth, changes in GDP imply changes in trading patterns between A and N. Consequently, it is difficult to assure that CL(ΔR) indeed corresponds to the theoretical concept, as outlined above.

The concept of carbon leakage did not exist before the UNFCCC came into force (Grubb, 1990). Over time economic literature identified mechanisms which may result in carbon leakage. The main mechanisms underlying this phenomenon are threefold (Rauscher, 2005), (Burniaux and Oliveira Martins, 2012).

- Changes in prices of fossil fuels. An additional abatement effort in A results in lower demand for fossil fuels, which thus leads to their lower price. This, in turn, provides incentives for additional use of fossil fuels in N.
- Changes in prices of final goods. An additional abatement effort in A results in higher prices of carbon-intensive goods there. Their production moves to N, which lead to higher emissions of carbon dioxide in N.
- Changes in production factor prices. An additional abatement effort in A reduces the remuneration of capital there. The capital moves to N, which leads to higher emissions of carbon dioxide in N.

This is sometimes referred to as the "green paradox" (Sinn, 2008); the "greener" an economy of a country (or region) becomes, the "dirtier" the world becomes. Another paradox is the so-called rebound effect, first observed by Jevons (1865), who predicted that increased efficiency of a steam engine would not lower the demand for coal. The rebound effect reminds us that, despite the "greening" of certain operations, an economy might become "dirtier" as a result of increased demand for these operations.

Calculating carbon leakage requires using a complicated analytical tool called the Computable General Equilibrium (CGE) model. Such a model looks at an economy as a whole and takes into account all inter-linkages: not only direct ones, but also indirect ones, since relationships between various sectors or activities can be quite convoluted.

Even though the mechanism is impossible to ignore, economists dispute about its scale. One way to explain the differences obtained by various authors is to look at elasticity assumptions (i.e. assumptions on how one variable reacts to the change of another variable). Gerlagh and Kuik (2007) carried out a meta-analysis of the literature and estimated an econometric model which established a link between elasticity parameters and CL. However, their analysis cannot be considered quite satisfactory as they assumed a production function which – by definition – assumes the leakage rate to be lower than 100%. Thus, for instance, they did not include a study by Babiker (2005), which demonstrates a higher

CL rate (because increasing returns to scale are assumed for energy-intensive sectors). A similar effect, i.e. a higher rate of CL, is implied by the product homogeneity assumption, i.e. assuming that products originating from different countries are identical. It is thus not surprising that the combination of constant returns to scale and product differentiation (i.e. the assumption that products originating from different countries are not perfect substitutes) – commonly assumed in many studies – implies a low leakage rate. A typical pattern found in CGE literature is consistent with Carbone *et al.* (2009), where the leakage rate is around 50% when goods are homogenous across regions and over 20% otherwise.

Some authors demonstrate that CL rate depends on the number of countries that belong to the abating (A) region, but conclusions may differ. In particular, while an OECD simulation (Burniaux et al., 2009) shows that CL should not exceed 6% when the whole Annex I is included in A, Winchester (2011) obtains a 25% carbon leakage. Boehringer at al. (2010) found leakage to be 10% in the case of a unilateral US policy, 25% in the case of a unilateral EU policy, and 15% if the EU and the US undertake a joint action.

A study by Kiuila *et al.* (2014) found that CL depends crucially on various parameters assumed, and – in particular – on so-called Armington elasticities. The latter reflect the fact that availability of an inexpensive import does not necessarily lower the demand for domestic production. While this has been observed empirically for many years, many on-going processes – such as lowering transportation costs and lowering tariffs – may modify Armington elasticities towards more homogeneous products. Unfortunately, there is little empirical research to support alternative assumptions, and yet they may change the CL rate drastically. Kiuila *et al.* (2014) demonstrate that CL is lower than 100% in most cases, but its instability is a matter of concern.

Troubled with CL, policy makers have analysed anti-leakage instruments. These include, for instance, Border Tax Adjustments (BTA), a sort of a trade sanction, mentioned above in 3.1.3. A BTA requires that a good imported from a region which does not abate to the same extent as the importing country is subject to a special tariff. Obstacles to this instrument are both practical and legal. The former stem from the fact that it is extremely difficult to calculate what part of the imported good should be subject to an extra payment, as its components are likely to originate from different countries. The latter simply means that it is difficult to reconcile BTAs with the no-trade-sanction principle adopted by the WTO. Some lawyers argue that BTAs can be reconciled with free trade rules (Hufbauer *et al.*, 2009), but most of the observed international developments demonstrate that this is questionable. Kiuila *et al.* (2014) add to this that such anti-leakage instruments are likely to decrease welfare for all concerned – both "rich" and "poor".

Enthusiasts of unilateral abatement criticise CL analyses based on CGE models as hypothetical, since obviously every CGE model is highly theoretical and its predictions are difficult to validate. A more convincing method would thus be to calculate CL empirically, but there have been very few attempts to do this. Aichele and Felbmayr (2012) is a rare example of an empirical approach. The authors show that ratification of the Kyoto protocol reduces a country's carbon footprint of production, i.e. the amount of carbon dioxide emission implied by production, but not carbon footprint of consumption. Thus leakage is 100%. A similar conclusion was arrived at by studying the impact of the Kyoto Protocol on the British carbon footprint (Committee 2013).

Environmental economics literature leaves no doubt that CL is a serious phenomenon which cannot be overcome but by seeking a global agreement on capping carbon dioxide emissions. Unilateral abatement does not help the problem and can be even harmful (Ritter and Schopf, 2014).

3.2.5 Will co-benefits help?

To a large extent carbon dioxide abatement implies lower fossil fuel combustion. Burning fossil fuels results not only in carbon dioxide emissions, but also in a number of other externalities, such as acidification, particulate matter emission, noise, and landscape disruption. Hence low carbon scenarios are likely to improve the general environmental predicament as well. Some analysts claim that such co-benefits (as they are often called) can overcome the public good problem: sovereign countries will adopt low carbon policies not necessarily for the sake of climate protection, but for the sake of "private" benefits, i.e. benefits that are to be reaped locally.

The argument is sound, and there have been numerous attempts to persuade non-Annex I countries to abate for the sake of immediate e.g. health benefits. Many non-Annex I country governments are economically rational, so it will be interesting to see whether this argument can be relied on. Unfortunately its relevance is limited. Based on European Union research projects, Zylicz and Czajkowski (2009) analysed the difference between externalities caused by fossil fuel combustion and by producing energy from renewable sources.

Quantified marginal costs of electricity production in EU27 in 2005–2010 [€/MWh]

	Nuclear power plant	**Hard coal condensing power plant**	**Lignite condensing power plant**	**Natural gas, gas turbine**	**H-power (dams and reservoirs)**	**Wind-mills, on-shore**	**Solar PV, open space**
Total External Cost	5.32	47.93	48.15	22.83	2.30	1.33	11.69
Human + Environmental	4.72	12.35	9.67	4.44	1.61	0.92	7.05
GHG	0.60	35.58	38.48	18.39	0.68	0.41	4.65
Total Private Cost	29.33	32.58	26.83	47.38	73.50	61.10	506.53
Total Social Cost	**34.65**	**80.51**	**74.98**	**70.21**	**75.80**	**62.43**	**518.22**

Source: Zylicz and Czajkowski (2009)

These numbers can be interpreted in two ways. First of all, the "Total External Cost" row gives Pigouvian tax rates (theoretical "penalties") that should be applied to all respective energy sources, if externalities are to be taken into account. Alternatively, if (e.g. for political reasons) the most environmentally disruptive source – i.e. lignite – is to be left untaxed, the other sources deserve subsidies to account for their relative environmental friendliness. The level of a subsidy can be calculated by subtracting a respective marginal cost from the largest number, i.e. €48/MWh. The numbers quoted for EU27 may be considered relatively low. This reflects the fact that European power plants are subject to fairly strict environmental regulations (see chapter 2.2), so that the remaining pollution is not as harmful as it used to be there in the 20th century, nor as it is in some developing countries now. If the table was to be replicated for many other countries, the external costs would have been higher.

Looking at the differences between climate related damages and the totals, one sees that the former account for a major part of the latter. This result depends mainly on the somewhat arbitrarily assumed valuation of cumulative climate-related damages at €21 per tonne of CO_2-equivalent emitted to the atmosphere. However, even if the most conservative estimates are adopted – e.g. reported by Nordhaus (2007) or Tol (2005), where damages of 1 tonne of CO_2-equivalent emitted are roughly a half of the ones presented above – the GHG emissions constitute by far a major part of the total external costs. As a result, only very modest climate protection policies can count on environmental co-benefits to be an important motive to "de-carbonize" the economy. If one were to implement a Pigouvian tax at the level of marginal GHGs emissions damages, the resulting improvement in health, crops, biodiversity and materials conditions caused by a switch from fossil to renewable technologies could only partly justify such a policy. However, if the global climate is to be protected more aggressively, then environmental co-benefits become relatively unimportant and they cannot be expected to motivate a substantial cut on fossil fuel use.

Chinese projections on future emissions with or without taking into account environmental co-benefits are illuminating. Indeed, co-benefits slow down the growth of CO_2 emissions, but they cannot reverse the trend even in a fifty-year perspective: without co-benefits emissions are expected to more than double, while with co-benefits they will grow by 50% "only" (Hu, 2004). In other words, even though climate protection is helped a little bit if co-benefits of climate policies are taken into account, it cannot be achieved.

The predicament can be different in developing countries, which have not implemented strict measures to cut down on "local" emissions. For this reason environmental co-benefits might play a more important role there than in the EU27. For instance, Dudek *et al.* (2001) estimate that, in Russia, ancillary benefits of an aggressive greenhouse gas reduction policy could cut the health risk in half, saving around 35,000 lives each year. Using conservative valuation procedures and applying high discount rates, this implies the benefit of $60 attached to each tonne of CO_2 abated. Here ancillary benefits seem to be a decisive factor supporting environmental protection. This result, however, is probably due to the very low initial level of environmental protection in Russia, which makes immediate improvements fairly obvious. It can be expected in some transitional economies as well, but benefits will soon become more difficult to demonstrate once the potential for easy solutions has been exhausted. In a longer-term perspective these economies will find it costly to reconcile global considerations with immediate environmental benefits. In addition, there probably exist far more cost-efficient ways of cutting down particulate matter and sulphur dioxide emissions, and thus saving lives, than switching to renewable technologies. This finding further undermines environmental co-benefits as a justification for GHG reduction policies.

Finally, a glimpse at "learning-by-doing" benefits of promoting renewable energy sources shows that, even though this effect may be quite substantial for some current technologies (mostly PV, i.e. photovoltaics), it remains too small to provide a meaningful incentive for substituting carbon and nuclear technologies with renewable sources. The same study (Zylicz and Czajkowski, 2009) looked at the expected level of private and external marginal costs of electricity generation in 2030, taking into account EU targets for a share of renewable energy in total energy produced. It showed that private cost of the cheapest (and thus most efficient) renewable technologies today – hydro and wind power plants – will not change substantially. The only significant change may be achieved in solar photovoltaic cost, which will, however, still remain the most expensive one, and as a result, will still not be a feasible major alternative. Since direct solar energy will not be cost effective, probably it will not have an important share in energy generation mix, and thus the result of learning-by-doing will be limited.

A breakthrough, however, can be brought about by an institutional change expected as a result of "prosumerism" in electricity production. Electricity consumers are expected to deploy new technologies in order to reduce the demand for energy from traditional sources. Decentralisation of electricity production may trigger technological progress faster than expected.

Negative co-benefits, i.e. damage caused by attempts to abate carbon dioxide, should be noted as well. For instance, a requirement to substitute traditional light bulbs with more energy efficient ones has promoted the production and use of fluorescent bulbs. These, however, imply a (small) mercury emission, which can be almost eliminated if the electric waste is handled properly. But 100% perfection is impossible. Thus – in this particular case – carbon dioxide abatement provides costs, and not only co-benefits. Likewise, ecologists complain that wind power is not bird-friendly, and so on. Co-benefits surely exist, but their quantification requires taking into account a wider context.

3.3 Biodiversity preservation

Two important conventions were presented at the 1992 Earth Summit in Rio de Janeiro: the UNFCCC, which was discussed above in 3.2, and the Convention on Biological Diversity (CBD). Abbreviated to "biodiversity", the name substituted for the older concept of nature protection.

According to ecologists, there are three layers of biodiversity. The most obvious one is species diversity. If a species becomes extinct, a loss is apparent. Somewhat less obvious is the concept of genetic or population diversity. Within a single species there may be many genetic traits (or many populations, each equipped with a similar genotype). This diversity may enhance the resilience of a species when perturbations – e.g. pest infestation – happen. On top of that, scientists discern landscape diversity, and additional characteristics of the living world, containing specific information on energy flows, trophic chains and other relations between animals and plants. In other words, even if all the species and their genotypes were preserved in zoological and botanical gardens, not all the information would be available to researchers and the public at large. Biodiversity is thus the diversity observed in the living world on different layers.

In the 1980s, it became clear that traditional nature conservation measures – focused on national parks and species protection – were insufficient to allow future generations to benefit from the Earth's living resources. As the preamble to the convention stated, there was a "general lack of information and knowledge regarding biological diversity", and at the same time there was an "urgent need to develop scientific, technical and institutional capacities to provide the basic understanding upon which to plan and implement appropriate measures". The convention called upon all states – which "have sovereign rights over their own biological resources" – to cooperate in order to "conserve and sustainably use biological diversity for the benefit of present and future generations".

Even though biodiversity is not confined to species protection, concerns about the fate of endangered animals and plants played the role of precursors of the CBD. They materialized much earlier, in 1973, when the Washington Convention (the Convention on International Trade in Endangered Species, CITES) was signed. The CITES included trade sanctions, though they are not as tough as the ones envisaged by the Montreal Protocol. Parties to the convention may trade with non-signatories, but they have to comply with certain principles. Participation has been almost universal, and the convention did contribute to nature conservation. Nevertheless, environmentalists realized that a truly effective instrument requires protecting entire habitats rather than specific animals or plants.

3.3.1 Biodiversity as a public good

The first thing that an economist has to clarify is whether biodiversity is a private or a public good. When it is private, no convention is required; a sufficient level of protection could be achieved at the national level. There are numerous examples of immediate local benefits derived from preserving a species or habitat. Farmers in Costa Rica sell – as souvenirs for tourists – beautiful butterflies that they can get for free in the tropical forest nearby. Many fishing ponds are carefully protected by local populations in order to benefit from the natural resources contained there.

Nevertheless, the benefits that come from biodiversity transcend the local scale in a spectacular way. An interesting case was provided by a discovery in a Mexican village of an endemic perennial grass that is closely related to corn (which requires ploughing every year). If the genetic information included in this grass allowed for developing a perennial corn, the benefits of an easier cultivation (and lower erosion) were estimated at $4 billion per annum in the United States alone (Fisher and Krutilla, 1985).

Biodiversity complies with the definition of a public good. The non-rivalry principle refers to the fact that many simultaneous users can benefit from the same information. The non-exclusion principle means that even those who do not contribute to biodiversity preservation can benefit from the information preserved.

As in the case of any public goods, incentives to free-ride result in their insufficient supply. In other words, the level of biodiversity protection is lower than justified by its total benefits. Every country protects nature to the extent that it views it as beneficial for its own purposes, regardless of gains that may accrue to other countries. That is why an international convention was called for.

3.3.2 The 1992 convention on biological diversity

Drafting the convention was a long process that culminated in several conferences carried out in United Nations Environmental Programme headquarters in Nairobi. The agreed text was presented at the Rio Earth Summit. It came into force in 1993, three months after the 30th ratification. With 193 parties, it is almost universally accepted.

As with many other environmental agreements, the convention has been criticised for its "toothlessness". Indeed it does not have bite, but it would be very difficult to establish quantified targets, such as in the case of Geneva LRTAP, Helsinki Baltic, Vienna Ozone, and the Rio Climate Convention. In the 1990s the European Commission ordered preparation of a report on Prominent European Environmental Problems (PEEP). The report identified 14 PEEPs, such as e.g. acidification, climate, noise, etc. It estimated costs of hypothetical improvements. There was, however, one PEEP doomed to worsen no matter what, i.e. despite any conceivable protective measures. This was biodiversity (RIVM, 1999).

In other words, even though biodiversity can be measured and analysed formally, it would be impractical to call for specific targets to which signatories should commit. Instead, the CBD has developed modalities to create processes and institutions important for biodiversity protection. The so-called *Aichi biodiversity targets* (adopted at the 10th Conference of Parties to the CBD in Nagoya in 2010) serve as perhaps the most comprehensive list. It delineates twenty targets grouped under five strategic goals (CBD, 2010).

> <u>Strategic Goal A</u>: *Address the underlying causes of biodiversity loss by mainstreaming biodiversity across government and society*

1. *By 2020, at the latest, people are aware of the values of biodiversity and the steps they can take to conserve and use it sustainably.*
2. *By 2020, at the latest, biodiversity values have been integrated into national and local development and poverty reduction strategies and planning processes and are being incorporated into national accounting, as appropriate, and reporting systems.*
3. *By 2020, at the latest, incentives, including subsidies, harmful to biodiversity are eliminated, phased out or reformed in order to minimize or avoid negative impacts, and positive incentives for the conservation and sustainable use of biodiversity are developed and applied, consistent and in harmony with the Convention and other relevant international obligations, taking into account national socio economic conditions.*
4. *By 2020, at the latest, Governments, business and stakeholders at all levels have taken steps to achieve or have implemented plans for sustainable production and consumption and have kept the impacts of use of natural resources well within safe ecological limits.*

<u>Strategic Goal B</u>: *Reduce the direct pressures on biodiversity and promote sustainable use*

5. *By 2020, the rate of loss of all natural habitats, including forests, is at least halved and where feasible brought close to zero, and degradation and fragmentation is significantly reduced.*
6. *By 2020 all fish and invertebrate stocks and aquatic plants are managed and harvested sustainably, legally and applying ecosystem based approaches, so that overfishing is avoided, recovery plans and measures are in place for all depleted species, fisheries have no significant adverse impacts on threatened species and vulnerable ecosystems and the impacts of fisheries on stocks, species and ecosystems are within safe ecological limits.*
7. *By 2020 areas under agriculture, aquaculture and forestry are managed sustainably, ensuring conservation of biodiversity.*
8. *By 2020, pollution, including from excess nutrients, has been brought to levels that are not detrimental to ecosystem function and biodiversity.*
9. *By 2020, invasive alien species and pathways are identified and prioritized, priority species are controlled or eradicated, and measures are in place to manage pathways to prevent their introduction and establishment.*
10. *By 2015, the multiple anthropogenic pressures on coral reefs, and other vulnerable ecosystems impacted by climate change or ocean acidification are minimized, so as to maintain their integrity and functioning.*

<u>Strategic Goal C</u>: *To improve the status of biodiversity by safeguarding ecosystems, species and genetic diversity*

11. *By 2020, at least 17 per cent of terrestrial and inland water, and 10 per cent of coastal and marine areas, especially areas of particular importance for biodiversity and ecosystem services, are conserved through effectively and equitably managed, ecologically representative and well connected systems of protected areas and other effective area-based conservation measures, and integrated into the wider landscapes and seascapes.*

12. *By 2020 the extinction of known threatened species has been prevented and their conservation status, particularly of those most in decline, has been improved and sustained.*
13. *By 2020, the genetic diversity of cultivated plants and farmed and domesticated animals and of wild relatives, including other socio-economically as well as culturally valuable species, is maintained, and strategies have been developed and implemented for minimizing genetic erosion and safeguarding their genetic diversity.*

<u>Strategic Goal D</u>: *Enhance the benefits to all from biodiversity and ecosystem services*

14. *By 2020, ecosystems that provide essential services, including services related to water, and contribute to health, livelihoods and well-being, are restored and safeguarded, taking into account the needs of women, indigenous and local communities, and the poor and vulnerable.*
15. *By 2020, ecosystem resilience and the contribution of biodiversity to carbon stocks has been enhanced, through conservation and restoration, including restoration of at least 15 per cent of degraded ecosystems, thereby contributing to climate change mitigation and adaptation and to combating desertification.*
16. *By 2015, the Nagoya Protocol on Access to Genetic Resources and the Fair and Equitable Sharing of Benefits Arising from their Utilization is in force and operational, consistent with national legislation.*

<u>Strategic Goal E</u>: *Enhance implementation through participatory planning, knowledge management and capacity building*

17. *By 2015 each Party has developed, adopted as a policy instrument, and has commenced implementing an effective, participatory and updated national biodiversity strategy and action plan.*
18. *By 2020, the traditional knowledge, innovations and practices of indigenous and local communities relevant for the conservation and sustainable use of biodiversity, and their customary use of biological resources, are respected, subject to national legislation and relevant international obligations, and fully integrated and reflected in the implementation of the Convention with the full and effective participation of indigenous and local communities, at all relevant levels.*
19. *By 2020, knowledge, the science base and technologies relating to biodiversity, its values, functioning, status and trends, and the consequences of its loss, are improved, widely shared and transferred, and applied.*
20. *By 2020, at the latest, the mobilization of financial resources for effectively implementing the Strategic Plan for Biodiversity 2011–2020 from all sources, and in accordance with the consolidated and agreed process in the Strategy for Resource Mobilization, should increase substantially from the current levels. This target will be subject to changes contingent to resource needs assessments to be developed and reported by Parties.*

As can be seen from this list, the Convention avoids setting specific, quantified targets, unless their interpretation is "soft" anyway (e.g. targets 5, 11, and 15). Some targets acknowledge the fact that – despite efforts – the predicament will worsen (e.g. targets 5, 9, 10, and 13). Nevertheless, bureaucratic actions (such as targets 2 and 4) should not be discredited. It may take many years before understanding of biodiversity becomes so universal that governments feel compelled to take it seriously, like they do when it comes to air protection.

The Nagoya meeting also formalized the so-called ABS, i.e. *access and benefit sharing*, a financial mechanism to achieve efficiency in providing biodiversity. In plain language, ABS is about paying for access to genetic resources. Economic efficiency requires that those who provide genetic resources or protect biodiversity as a source of genetic information should be remunerated. The convention recognized this fact from the outset. Its article 15.7 states that access to genetic resources is organized by the parties "on mutually agreed terms". A typical situation is that a "rich" country firm, say, a pharmaceutical company, collects herbs in a "poor" tropical country. An herb may then be used in order to develop a medical drug or a cosmetic leading to a market success. The "poor" country, or – more specifically – a community who stewarded that piece of wildlife, should have some share in commercial benefits.

There is no reporting on article 15.7, but some transactions have been documented in environmental economics literature, at least partially. It is difficult to estimate their total scale, so a well-publicized transaction between Merck & Co. (a multinational pharmaceutical company) and INBio, the National Biodiversity Institute of Costa Rica, will serve to illustrate the argument here (Lesser and Krattiger, 2007). In the 1990s Merck paid $ 1.1 million to INBio for bioprospecting and agreed to pay an undisclosed royalty for resulting product sales. The payment was in the form of training services for INBio staff and equipment purchases. This was considered a fulfilment of article 15.7 of the convention, and hence acquiring the right to explore local biodiversity.

The Merck transaction is an occasion to discuss several issues linked to ABS. First of all, the convention is binding for governments, not for firms. Therefore the pharmaceutical company paid something not because it was directly bound to do so, but because its government could (and should) require that biodiversity prospecting is carried out without violating article 15.7. It is not clear who is the payee. One way of complying with 15.7 would be to pay to the government whatever is "mutually agreed upon". However, this is not consistent with the spirit of the convention, which calls for remunerating those who steward the biodiversity, and some prospective payees are afraid that the government will not forward the payment as anticipated. Paying physical persons involved in collecting herbs or active in the biodiversity preservation does not seem appropriate either.

Therefore, in some transactions "rich" country firms paid local environmental NGOs, which had some obvious disadvantages too.

In addition, in the early 1990s some environmentalists were afraid that article 15.7 might facilitate species extinction, rather than preservation. A Machiavellian scenario was contemplated that a pharmaceutical firm collects an endemic sample, patents some crucial genetic information and destroys the *in-situ* resource in order to become a monopolist. Even though theoretically possible, this scenario has never materialized.

Perhaps a more realistic scenario is to agree on a royalty fee in order to permit patenting. Then the owner of the patented genetic information can change the DNA information slightly by means of genetic engineering, and patent the new pattern. This new pattern would be an unquestionable property of the laboratory rather than the joint property of the laboratory and whomever was the owner of the original resource. If a commercial product uses this modified information (even though true inspiration came from the original pattern) then the original royalty loses its value. There are suspicions that some transactions can be arranged along these lines, but the lack of reporting does not allow for the verification of this claim.

Finally, inspiration originating from biodiversity is sometimes difficult to quantify. In 1970, Du Pont patented Hollofil®, an insulating material consisting of very thin fibres, each of which was empty inside. The company announced proudly that inspiration came from observing polar bears whose hairs are empty inside. This property has two important implications. First, it lets the mass of the "insulation" layer be lower than for traditional furs. Second, it lets solar energy reach the skin of a beast easier and warm it more effectively on a sunny day. Should Du Pont pay anything to anybody who protects polar bears? Would a hypothetical extinction of polar bears make the inspiration which led to patenting Hollofil® disappear?

Ambiguities related to applying article 15.7 prompted CBD parties to negotiate the Nagoya Protocol (2010) devoted entirely to ABS. The protocol emphasizes that traditional knowledge of genetic resources (not only patented information) shall be remunerated. It establishes a number of mechanisms to ensure that this knowledge is not overshadowed by tricks aimed at making commercial laboratories turn out to be the main source of inspiration. It is difficult to ascertain whether this new protocol becomes successful at providing right incentives for biodiversity preservation. The protocol compiled fairly exhaustive lists of benefits. They read:

> 1. *Monetary benefits may include, but not be limited to:*
>   (a) *Access fees/fee per sample collected or otherwise acquired;*
>   (b) *Up-front payments;*

(c) Milestone payments;
(d) Payment of royalties;
(e) Licence fees in case of commercialization;
(f) Special fees to be paid to trust funds supporting conservation and sustainable use of biodiversity;
(g) Salaries and preferential terms where mutually agreed;
(h) Research funding;
(i) Joint ventures;
(j) Joint ownership of relevant intellectual property rights.

2. *Non-monetary benefits may include, but not be limited to:*
(a) Sharing of research and development results;
(b) Collaboration, cooperation and contribution in scientific research and development programmes, particularly biotechnological research activities, where possible in the Party providing genetic resources;
(c) Participation in product development;
(d) Collaboration, cooperation and contribution in education and training;
(e) Admittance to ex situ facilities of genetic resources and to databases;
(f) Transfer to the provider of the genetic resources of knowledge and technology under fair and most favourable terms, including on concessional and preferential terms where agreed, in particular, knowledge and technology that make use of genetic resources, including biotechnology, or that are relevant to the conservation and sustainable utilization of biological diversity;
(g) Strengthening capacities for technology transfer;
(h) Institutional capacity-building;
(i) Human and material resources to strengthen the capacities for the administration and enforcement of access regulations;
(j) Training related to genetic resources with the full participation of countries providing genetic resources, and where possible, in such countries;
(k) Access to scientific information relevant to conservation and sustainable use of biological diversity, including biological inventories and taxonomic studies;
(l) Contributions to the local economy;
(m) Research directed towards priority needs, such as health and food security, taking into account domestic uses of genetic resources in the Party providing genetic resources;
(n) Institutional and professional relationships that can arise from an access and benefit-sharing agreement and subsequent collaborative activities;
(o) Food and livelihood security benefits;
(p) Social recognition;
(q) Joint ownership of relevant intellectual property rights.

The CBD is not as widely publicized as UNFCCC, but it is an important international agreement too. Both of them inherited certain concepts from the Montreal Protocol, like the principle of "rich" countries paying for incremental costs in "poor" countries. However, unlike the UNFCCC, the CBD looks at the dichotomy of both categories in a more rational manner. The convention (as well as its subsequent protocols) introduces the category of "economies in transition" (art.20.2). They are not "poor", but they are not obliged to pay as the "rich". They can voluntarily accept certain obligations reserved for the latter.

It would be interesting to analyse why this solution was accepted in the CBD and not in UNFCCC. Until now, its practical implications have not been very important, as the costs of the CBD are fairly low. Unlike in the case of climate protection measures, biodiversity projects have rather small budgets. A typical cost of a species preservation programme is in the order of million rather billion euros. Nevertheless, the fact that the CBD has never introduced the concept of "common but differentiated responsibility" is likely to maintain the convention's effectiveness even when its level of ambition increases in the future.

4. Cross-cutting Issues

International environmental cooperation is not confined to specific environmental issues like transboundary pollution or biodiversity preservation. It is also implicit in many "horizontal" agreements adopted, or activities carried out, in order to address certain economic problems. However, they may be relevant for environmental issues too. That is why it is important to study such non-environmental phenomena.

4.1 International development assistance

The concept of "assistance" or "aid" is misleading. In common language it is understood as a gesture motivated by altruism. A much broader understanding is used in statistics. First of all, international statistics cover so-called development aid, i.e. money offered for economic development purposes. Charities are excluded, unless there is a large scale disaster, and the international community feels obliged to react. This is called "humanitarian aid". It typically accounts for much less than 10% of the total (only in Africa does it occasionally get closer to 10%).

Virtually all of aid money originates from OECD countries. The most wealthy of them constitute the Development Assistance Committee (DAC), a club of 28 countries and the European Union. Hence DAC statistics are commonly referred to in development aid analyses. DAC numbers outweigh anything else by a wide margin. For instance, one of the largest agendas in the UN system, the United Nations Development Programme (UNDP), disburses just several percent of what is captured by DAC statistics.

Adopted by DAC, definitions of development assistance is quite formal. Official Development Assistance (ODA) is defined as those resource flows to countries and territories on the DAC list of ODA recipients and to multilateral development institutions which meet the following criteria (OECD, 2008):

(a) they are undertaken by the official sector (rather than private entities);
(b) with promotion of economic development and welfare as the main objective (even though they may serve other purposes as well);
(c) at concessional financial terms (in the case of a loan, the pay-back terms should be more favourable than the standard 10% interest, and no grace period; a grant element of these terms should be equivalent to at least 25%);
(d) grants, loans and credits are not for military purposes.

As seen from this definition, ODA excludes certain flows that could have been considered "assistance" in common language. On the other hand, the third criterion is quite liberal. For instance, a loan whose payback term is 7% instead of 10% meets the (c) requirement. Hence, numbers quoted as ODA may not fully reflect what is deemed as "assistance", but there are no better statistics.

The standard figure for ODA displays a large discrepancy in terms of its share in the Gross National Income (GNI) of the donor countries. The United Nations established a target of 0.7% GNI to be transferred from the "rich" to the "poor". Initially, only Norway complied with this threshold. In 2013 there were five countries with ODA corresponding to more than 0.7% of their incomes. Most countries are far below it. It is striking that – in relative terms – the United States is below 0.2%, i.e. a half of the index for Germany. However, the index may be somewhat misleading. In Europe people rely on their governments to a large extent. If they feel something should be done, they expect the government to do it (since it taxes them considerably). In contrast, Americans do not expect their government to do everything, and if they feel something should be done, they often do it through non-government organizations. Apparently this pattern applies to assistance activities too. In addition to ODA (which covers the official sector only), there are separate statistics for grants offered through "private voluntary agencies". They account for roughly $30 billion per annum, and they are by far dominated by American contributions (over $23 billion). When this information is taken into account, the US emerges as a key international donor: the largest one in absolute terms, and also a significant one in relative terms.

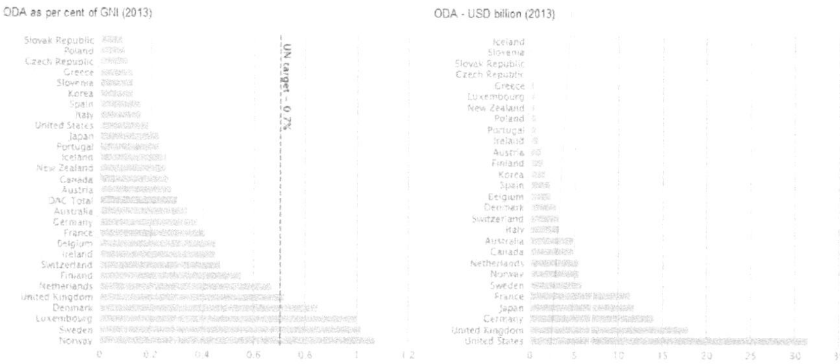

Source: OECD (2014)

The trend for ODA is not spectacular. Since the 1970s, ODA as percent of GNI oscillates between 0.2% and 0.4%. After the year 2000 it increased somewhat, but the growth is far from steady. The same index for Poland (after 1997) is even less spectacular, but it has had an increasing trend since the accession to the European Union.

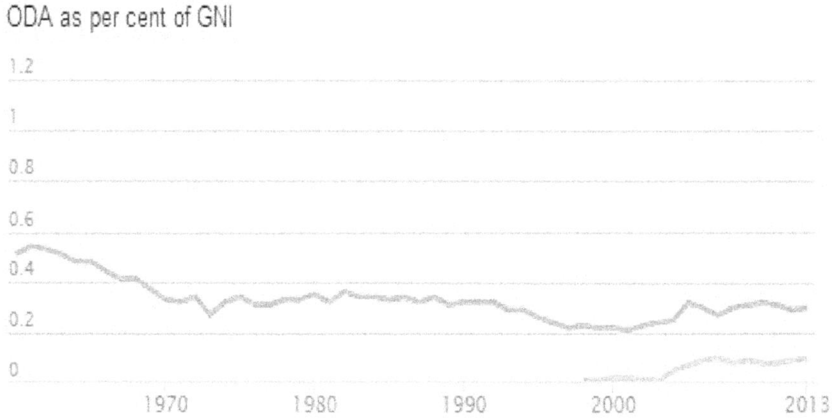

Source: OECD (2014)

The total ODA was almost $180 billion in 2012. In current prices, it has grown by 3%-4% over the last couple of years. Given inflation, this growth has been slower in real terms. In per capita terms, it can be related either to the donor or beneficiary populations. The "rich" countries account for some 1 billion people. The "poor" countries account for some 6 billion people. Hence, the total of $180 billion corresponds to $180 per head in donor countries and $30 per head for an average recipient of ODA. There are, however, large geographical and time variations. As a rule, countries that attract political attention receive more in relative terms. Iraq received over $20 billion in 2005, but between 2009 and 2012 it got less than $2 billion per annum. In contrast, between 2009 and 2012, Afghanistan received more than $ 5billion each year, while between 2003 and 2008 it was less than $2.5 billion.

It is difficult to identify environmental components in these numbers. OECD statistics identify such categories as "Water and sanitation", "Energy", and "Agriculture, Forestry, Fishing". These are certainly broader than environmental protection, but their total may serve as its upper estimate. This total grew from $15.5 billion in 2006 to $35.7 billion in 2012, i.e. from 12% to 20% (OECD, 2014). It can be assumed that environmental protection has absorbed 10%-15% of ODA.

Much more detailed information is available for Poland, which attracted the attention of international donors in the 1990s. Environmental assistance to

Poland received much publicity. On the one hand, taxpayers in many developed countries were inspired by news about environmental disruption in the former Soviet bloc. On the other hand, Poland offered a fairly stable administrative infrastructure, promising much less corruption than in alternative destinations. As a result, Poland received quite sizable environmental assistance. The amount pledged for 1991–1996 was $230.1 million, i.e. $46 million per annum – slightly over $1 per head. The world average ODA per head was higher, but Poland was certainly much richer than an average beneficiary in developing countries. Thus, many activists in Third World countries considered the donors' involvement in Poland as extravagant.

Even though environmental assistance for Poland was fairly high, it never matched domestic investment environmental expenditures. It accounted for 1%–2% in the early 1990s. The total foreign investment increased, but – to a growing extent – it was carried out commercially rather than as an ODA. Out of the assistance money in 1991–1996, 39% was spent on air protection, and 26% on water protection. An interesting feature of this process was that investment expenditures accounted for 57% of the total, while pre-investment ones (e.g. technical projects, analyses, etc.) – for 22%, and other (such as training) – 21% (GUS, 1997).

Distribution of expenditures was a matter of disagreement. Domestic experts claimed that only investment outlays contributed to environmental protection, while everything else, especially training, was a waste of resources. On the contrary, international experts insisted that unless a serious analytical effort is undertaken, investment money can be wasted. This controversy is typical for any international assistance, and it cannot be resolved easily. Indeed international experts have a natural preference for analyses and trainings, and domestic ones do not feel any need for outsiders looking into what seems to be obvious. Perhaps the proportion of assistance spent on investment could have been higher, but tangible savings were often achieved by reconsidering some technological choices or scaling down certain projects.

Yet another important characteristic of international assistance has to be noted. This is the issue of "tying". There is a natural tendency of donors to "tie" the aid, that is to demand that their money is spent on purchases from the donor country. By the way, often this is how taxpayers can be convinced to allocate a part of the budget for external assistance. Studies show that narrowing the market where the money is to be spent increases the cost of purchases by 15%–30%. An additional disadvantage of "tying" aid is the increased cost of bureaucracy both in the donor and recipient countries. In order to avoid this, in 2001 OECD issued a *Recommendation on Untying ODA* (to the so-called Least Developed Countries), which started to improve the situation. Whereas the share of untied assistance was once 46%, by 2008 it grew to 82%.

Aid flowing to Poland in the 1990s was by and large "tied". As a result, project costs were probably higher than necessary. Another outcome was excessive maintenance costs. For instance, environmental inspectors were happy to receive free monitoring equipment. Various regions benefitted from aid originating from different countries. When the tools required some maintenance or repair, it turned out that none of the manufacturers established local service centres (due to small markets for their respective products), and the equipment had to be shipped abroad. This could have been avoided if inspectors were given freedom to choose unified standards for their tools.

Substantial resources are channelled through international assistance programmes. Environmental protection projects account for 10%-15% of these flows. Occasionally they may be important for solving some problems, but they do not seem to address them in a systematic way. They may epitomize some sort of logic of international cooperation in general, but not international environmental cooperation.

4.2 The institutional framework for environmental protection

4.2.1 Debt-for-Nature swaps

Indebtedness and natural capital destruction are ranked among the most acute global problems affecting the present and future well-being of a large part of the world. First proposed by Lovejoy (1984), debt-for-environment swaps were once considered a wonderful tool to address the two problems simultaneously.

The first swap was implemented in Bolivia in 1987. Despite the publicity this precedent-setting agreement received in the media (Shabecoff, 1987), and despite the significant scale of the environmental improvement involved, the volume of the transaction was small. $650,000 of Bolivia's debt (a fraction of a percent of the country's substantial indebtedness) was bought for $100,000 (roughly 15 cents on the dollar) in the secondary financial market and retired. In exchange, the Bolivian government agreed to expand protected areas around the Beni Biosphere Reserve by 1.5 million hectares. It contributed $100,000 to the protection programme, which also received a $150,000 grant from the US Agency for International Development (Cody, 1988).

Not much changed in subsequent years. To the frustration of many environmentalists, despite a potential envisaged by some economists (Hansen 1989), debt-for-environment swaps (including so-called debt-for-development swaps, which cover environmental projects but may also address health care, education, etc.) accounted for only 1%–2% of approximately $38 billion channelled through debt-for-equity swaps (UNCTAD, 1992). The latter focused the attention of financial analysts, politicians, and international investors almost exclusively.

Moreover, some politicians and analysts have concluded that debt-for-environment swaps are not an adequate instrument for environmental management. This assertion is based on two premises. First, there may be no correlation between bilateral indebtedness and the creditor's environmental interest in its debtor country. Second, a debt swap is a very indirect, and thus probably inefficient way to subsidise environmental protection; a direct side-payment would be a more straightforward method to proceed. Whilst either of the arguments has its own merit, they can be questioned on purely theoretical grounds. They can also be questioned on practical grounds as prematurely rejecting an instrument which – when properly fine-tuned and targeted – may significantly improve countries' ability to manage the global commons.

Quite different reservations have been voiced by some environmentalists, who realised that most of the swaps concluded by the early 1990s did not result in any transfer of resources from the rich to the poor. What happened was that a "First World" environmental group gave money to a "First World" commercial bank in exchange for a "Third World" government decision to give money to one of its environmental groups (Mahony, 1992). Thus, at best, debt swaps helped to avoid some of the current net wealth transfers from South to North which would otherwise have taken place. This implies a question, namely whether it is worthwhile for scarce environmental budgets – both public and private – to contribute to an endeavour which only results in assisting Northern banks to improve their current balances, as well as to increase the likelihood of recovering the rest of their money in the future.

This scepticism was justified. Indeed, one could have easily given up with debt-for-environment swaps, if they were to be arranged on an *ad hoc* basis, and if they were not to be a part of a complex policy to solve the indebtedness problem, revert the net current flow of wealth, and create local institutions to manage the global commons. Zylicz (1992) discussed the institutional dimension of a debt-for-environment swap to conclude that a swap facility might be in a unique position to serve certain international environmental purposes.

It could be argued on theoretical grounds that there is no correlation between indebtedness and the environmental interests of a creditor in its debtor. The "no-correlation" argument was actually applied by one of Western European

Environment Ministry officials in 1991 to justify their government's lack of support for the Polish proposal to swap a portion of the debt. However, the notion of *an environmental interest* is a broad concept, and deserves to be examined. Country *A* can be said to have an environmental interest in country *B* under several alternative circumstances, listed below.

1. *B* is an "upstream" (or "upwind") polluter affecting *A*;
2. *B* is a provider of a public good whose supply is of interest to *A*;
3. *A* and *B* exploit the same common resource stock;
4. *A* and *B* consume services provided by the same common environment, but the value attached to these by *A* is higher than that attached by *B*;
5. *A* appreciates the option, vicarious, or existence value of a resource owned by *B*.

Indeed, it would be unreasonable to expect to find too many examples of the type (1) correlation. With creditor countries clustered in the North, and debtors in the tropics, there are little, if any, chances that a debtor may also happen to be a significant polluter affecting its creditor. Poland *vis à vis* Sweden may provide such a rare example. However, in the case of, for instance, Germany and Poland, the former (i.e. the creditor) is the upwind party. The type (2) correlation, best exemplified by present concerns to preserve global biodiversity, is commonly observed. Despite the establishment of exclusive economic zones, harvest quotas, etc., many renewable resources still possess typical characteristics of a commons. One can think of global warming in terms of carbon dioxide emissions which originate mostly from the North now, but whose long-term trends depend on development and energy policies carried out in the South. Hence, the likelihood of observing a correlation between indebtedness and the type (3) interest is definitely higher than under (1). Ozone layer depletion provides an example of a situation where "the North" (including Australia) attaches a much higher value to ozone protecting measures than the rest of the world and, therefore, it is interested in vigorously stimulating such measures elsewhere. This can be viewed as the type (4) relationship. Lastly, there exists wide interest in assisting low income countries in their struggle with poverty and hunger. Also, there exists a recognition of the fact that this struggle will never succeed unless the hungry and the poor prevent their natural capital stocks from further decline. Thus the North's willingness to assist – if materialised consistently – must manifest itself in its appreciation of vicarious values derived from those distant environments and resources. This explains much of the potential type (5) interests other than, say, the concern with biodiversity, which could also be classified under (2).

In sum, to the extent that the indebtedness problem affects first of all those:

- who control biodiversity (type 2),
- who can offer inexpensive carbon sinks (type 3),
- whose production and development plans do not undergo the same domestic checks as in the North (type 4), and
- whose survival very much depends on saving their resource base (type 5), its correlation with environmental interests is, in fact, implied.

To be precise, in order to claim the correlation between bilateral indebtedness and environmental interests, one needs an additional assumption that the volume of credit activities in the past (leading to present indebtedness) was roughly proportional to the size of the creditor's economic activities, the latter being indicative for its readiness to participate in global environmental programmes. At the same time, one should stress that these hypothetical interests cover global issues like biodiversity, climate change, ozone depletion, as well as maintaining debtor countries' natural resources rather than transboundary pollution externalities.

Nonetheless, even if environmental interests are related to indebtedness, then it could be questionable whether debt swaps are any better than direct side payments. Here the argument should address the preferences of both parties: creditor and debtor.

If a debt swap were simply a one-time subsidy to free the debtor of (a portion of) its current debt service in exchange for a domestic environmental expenditure, then certainly a direct subsidy to finance the expenditure would be a much more natural way to go. The idea of a swap, however, is to change the future flow of payments so as to assist the debtor in a more permanent manner. From the creditor's point of view such a deal may be perceived as an inflexible tool, as it requires a long-term commitment, typically of ten years or more depending on the time profile of the debt swapped. For the very same reason, from the debtor's point of view a swap is a more convenient form of assistance, since it secures a more stable source of funding than aid packages which typically have a much shorter time frame. It is assumed in this analysis that otherwise the debt would have been serviced. Here we do not address the question of a possible unilateral debt moratorium decision by the debtor. Such decisions free the economy from the immediate burden but simultaneously they cut it from international capital markets. Still, the debtor may hesitate to negotiate a swap, as long as there is a chance for an unconditional debt reduction. In fact this was the case in Poland before 1991. The government was involved in negotiating with the Paris Club an extensive debt-rescheduling package. The debt-for-environment swap initiative was launched only after the extent of negotiated unconditional debt reduction was perceived as final.

The considerations above suggest that creditors may have an inherent bias against swaps, while debtors may demonstrate an inherent preference for them, at least when they see no chance for reducing the debt burden otherwise. This leaves aside the fact that some debtors – e.g. Brazil – oppose debt-for-environment swaps as impairing their sovereignty over the national resource base. A declaration adopted at an international seminar hosted by the Brazilian Institute for Economic and Social Analysis (IBASE) states, among other things, that (Mahony, 1992):

> *The mechanism which exchanges foreign debt for environmental benefits does not contribute to the development of environmental policies which are consistent with democratic management of natural resources, and which might lead to environmental conservation and a better quality of life for the local population. Instead, it forms part of a general strategy for reaffirming the creditors' political and economic domination over the debtors, within a development model which commercializes life in all its aspects.*

But this is not an argument against the swaps specifically, but rather against any form of sharing their control of resources, and it would equally apply to direct payments (tied to certain environmental requirements), should such an offer been made. An additional reason why creditors may prefer direct payments (especially in the form of conventional aid programs with tied procurement) is that these payments are for the most part recaptured by entities from the donor country who receive contracts. It is therefore quite natural that aid packages rather than swaps are more likely to win support by political constituencies in creditor countries.

This prevailing philosophy assumes that environmental protection, or sound resource management is a technical question which can be solved once the money is there. It overlooks the fact that protection and management are carried out by institutions. Moreover, if they are to be carried out on a self-sustained basis, these institutions should be local ones, run by local professionals, contributing to the emergence of local constituencies for environmental and resource stewardship.

Seen from this point of view, debt swaps are an interesting tool for three reasons. *First*, unlike direct payments coming in the form of aid-packages, swaps provide a long-term source of financing; thus they make it possible to design long-term programs aimed not only at fixing concrete technical problems, but also at establishing required institutions (Zylicz, 1992). *Second*, by raising environmental issues at the debtor country's government forum, swap negotiations elevate the status of environmental departments and make them partners to financial and industrial agencies. *Third*, swaps are more likely to achieve leverage in terms of money spent on the environment per dollar invested in debt reduction; this is evident if the swap takes the advantage of the difference between the face value of the debt, and its price at secondary markets, but even if the debtor's commitment corresponds to the actual payment made by the creditor, the leverage can result from the requirement that the expenditures of the former are additional to its long-term spending plan. All in all, there is a tremendous potential inherent in debt-for-environment swaps which could make this instrument for developed countries more attractive than direct side-payments.

There are two aspects of a successful design of a debt-for-environment swap: (1) its general layout, as determined by debt conversion agreements; and (2) specific institutional arrangements establishing a facility to administer converted funds. In what follows we will briefly discuss both aspects.

The crucial difference between debt-for-environment and debt-for-equity swaps is that, in the latter, the creditor can recapture its assets at least partially and, perhaps, with some delay. An example of recovering full face value of the debt is provided by the sale of ex-East-German claims on developing countries to commercial firms engaged in importing raw materials; Griffith-Jones (1992) reports on the case of Zambia. As a rule, however, the creditor captures only a part of the debt. It is therefore no surprise that 98–99% of all debt conversions are confined to this particular form. Whilst successful in solving some of the financial problems in some Latin American countries (especially in Chile (UNCTAD 1992)), debt-for-equity swaps cannot contribute to capacity building in the same way as outlined above. Thus the first principle that the design of a debt-for-environment swap should start with is a clear identification of the swap as an environmental one. It should be clearly determined that the discharge of the debtor government's debt service obligations is linked to its financing domestic projects (previously agreed upon with the creditor), and not to the creditor government satisfying its domestic financial institutions affected by the debt conversion. Convincing creditors to switch from their myopic preference to fix their financial problem to a more comprehensive approach is thus the first challenge that the architect of a debt-for-environment swap has to face.

An important issue needs to be clarified here, namely the additionality of the debtor's expenditures claimed as the fulfilment of its obligations under the swap. Even though it is conceivable that creditors may agree to honour any expenditures demonstrated by their debtor (both in the public and the private sector), a more typical requirement would be to ensure that more projects are undertaken as a result of the swap. There are many different ways to approach this problem, but the most practical one is the macroeconomic test, namely to require that the country demonstrates increased overall environmental expenditures with respect to a historical baseline.

Assuming that a certain number of creditors agree to a swap, the next step is to avoid establishing a series of bilateral facilities each administering funds released in a given swap. There is a natural tendency for creditor governments to prefer such bilateralism, as they feel in this way they can have a better grip over the resources that will be spent on their behalf. But it would be clearly inefficient for the debtor country to establish a whole array of swap facilities – each handling only a portion of the resources made available through conversions. An additional disadvantage of a bilateral set-up is that in such circumstances it would be extremely difficult to avoid procurement tied to the creditor country. This is, perhaps, the major factor explaining creditor governments' preference for bilateral schemes: if the face value of the debt cannot be recovered in the form of equity shares, then, at least, its firms should benefit from contracts. Studies show that tying procurement to any given geographical location may increase the cost of contracts by as much as 30 percent (see 4.1). For that reason, the administration of swapped funds should take the advantage of an international competitive tender (either open or limited), which of course is difficult to defend within the framework of a bilateral facility.

At the same time, however, financial interests of creditor countries should be recognised too. A compromise can be arrived at by adopting a *geographical distribution* rule. This rule – widely applied in European Union programmes – offers every participating donor a possibility to recapture (in the form of contracts awarded) its share (with some flexibility margin) of the effort.

The main difference between the geographical distribution and tied procurement principles is that, in the former, the contracts are <u>not</u> *a priori* assigned to any given location. Awarded on a competitive basis, the contracts are periodically reviewed, and their geographical distribution is assessed. If this turns out to be sufficiently close to the desired one, no corrective action is taken. Otherwise, administration of the programs may take one of two steps: (1) modify the "short lists" of potential bidders so as to constrain the number of firms from the countries "over-represented" in contracts; or (2) modify the selection of projects so as to offer a better chance to those firms which have not found an opportunity to show their specific competitive advantage so far. Even though this is not a perfect solution, geographical distribution of contracts is a much better alternative than tied procurement, and it has proven viable in the EU practice.

Assuming that the multilateral profile of a debt-for-environment facility has been successfully secured, the next issue is to find a viable institutional form for representing creditors' interests without impairing the debtor's sovereignty. A natural way to go is to establish a supervisory board with a collective minority representation of creditors, offering the rest of the seats to debtor country nationals carefully selected from government, NGOs, and academia.

An ideal way to take decisions regarding general policy, project selection, and spending would certainly be an unanimous vote. For obvious reasons, this cannot be adopted as a charter principle, even though it may be actually followed most of the time. Requiring a qualified majority vote for a decision to be valid, and ensuring that the debtor's party cannot achieve such a vote unilaterally is a solution which may let creditors accept their minority representation. It does not seem to be appropriate to provide for a full representation of creditors on the board (with, say, votes proportional to their commitments), since this would inflate the size of this organ and increase costs and bureaucracy substantially. There are many international-regional institutions (Paris Club, OECD, EU, EFTA, World Bank, EBRD, etc.) where an initial aggregation of creditor preferences can be achieved, and which can be invited to represent groups of creditors, especially those with minor stakes.

4.2.2 The Polish EcoFund experience

In 1991–92 the government of Poland launched a large scale debt-for-environment swap intended to ease the country's indebtedness problem and address a couple of global environmental problems at the same time. The swap was remarkable not only in terms of money (it aimed at more than $3 billion, 10 percent of the debt to the "Paris Club" countries), but – even more importantly – in terms of its careful design. Substantial effort was put into developing an arrangement complying with principles discussed in 4.2.1. In addition, by that time, Poland had gained some experience with bilateral environmental assistance and understood that the cost-effectiveness of spending crucially depends on the design of an appropriate facility. Consequently, effort was also put into avoiding deficiencies of traditional bilateral mechanisms.

In 1992 the Polish Minister of Environment established the EcoFund, a non-government organization (NGO) registered as a foundation under Polish law. It has managed all debt-for-environment swaps that have been negotiated by the government since 1991. In 1994 the Economic Commission for Europe of the United Nations (UN ECE, 1994) officially recognized the Polish EcoFund as "an innovative and promising pilot project" to serve as a pattern for other heavily indebted economies in transition. In 1997 the foundation was subject to two independent reviews: by the OECD (1998) and by KPMG (1997). While the former was undertaken by the OECD on its own initiative, the latter was commissioned by the international Council which supervised the EcoFund. Both reviews confirmed the outstanding performance of the EcoFund, the adequacy of its original design, and the timeliness of its mission. Despite that, the participation of creditor countries has been rather unimpressive. Two countries declared entering the swap soon after its launch in 1991. They were joined by a third one a year later. The next two countries – one of them "on a trial basis" – came in 1997. They were followed by a sixth one in 1998. Given the fact that the number of potential participants is seventeen, i.e. the entire "Paris Club", the response should be seen as lukewarm.

The idea of a debt-for-environment swap was not quite novel in Poland in 1991. In 1990 the World Wildlife Fund (WWF) converted $50,000 (purchased in a secondary market for $11,500) of the Polish debt into a Mazurian Lake conservation project. Also in 1990, a bilateral Polish-Finnish agreement was reached to swap a substantial portion of the official bilateral debt for a subsidy (up to 30 percent of the contract price) for Polish importers who buy Finnish environmental protection and energy saving equipment. The latter was a typical bilateral agreement with all disadvantages of tied procurement and of an *ad hoc* institutional structure. Initially, implementation of the agreement was slow due to bureaucratic and procedural failures on the Polish side. Eventually, however, it became clear that because of the lack of appropriate incentives (a result of the tied-procurement philosophy), potential projects are not competitive compared with their open-market alternatives even after the 30 percent budgetary subsidy for Polish importers. Hence, the project pipeline was virtually empty for more than a year.

In what follows, a detailed calendar of events is referred to in order to analyse this unique development in international environmental cooperation. It is based on a description included in Zylicz (2000).

In 1990–91 the government of Poland was negotiating the scale of the general debt relief untied to commitments to domestically spend on anything negotiated with the creditors. No swaps were discussed openly at that time. By the end of February 1991, it became apparent that this scale – with respect to the Paris-Club debt – would reach 50 percent. Based on that, in his *Memorandum* of March 5, 1991, Dr. Maciej Nowicki, Minister of Environment, outlined Poland's environmental protection programs of international relevance. These programs would allow Poland to apply (with respect to the remaining 50 percent of the debt) to redirect its debt service for environmental protection purposes.

In particular, four main problem areas were identified in the *Memorandum* as international priorities:

- Long-range transboundary air pollution abatement;
- Improved control over contamination and eutrophication of the Baltic Sea;
- Carbon dioxide (and other greenhouse gases) abatement mainly through promoting new energy efficient technologies; and
- Preservation of biological diversity through sound development assistance to regions with ecological systems of international importance.

This served as the starting point for further delineating the initiative.

On April 4, 1991, Prime Minister Jan Krzysztof Bielecki appointed an Interim Interministerial Committee in charge of the debt-for-environment swap chaired by a Deputy Minister of Finance. The Committee was responsible for establishing an environmental fund, the EcoFund, managing the selection, development, and financing of projects undertaken within the framework of future debt-for-environment swaps.

Subsequent activities were based on a April 21, 1991 decision of the Paris Club, which determined that additional reductions of the Polish debt – up to 10 percent of the initial amount, i.e. up to 20 percent of the original level reduced by a half – may be achieved in a series of voluntary bilateral agreements with creditors. Hence, if all of the creditors positively responded to this option, the EcoFund would have received more than $3 billion (in net present value terms as of April 1, 1991) to be disbursed over 18 years. The scale of the arrangement would be much larger than anything in the past. Taking into account the time profile of the debt, the EcoFund's annual budget would have started with $120 million, and would have reached about $400 million by the year 2000, when peak debt payments were due. These are significant figures, even though they are much less than what the Polish economy spends on the environment domestically. It was estimated that the EcoFund's budget would never exceed, say, 15 percent of the total domestic environmental expenditures (Republic, 1992c).

In order to reinforce the Polish initiative, Prime Minister Bielecki approached Ms. Gro Harlem Brundtland, the Prime Minister of Norway, who has also earned worldwide acclaim in environmental matters (as the chairperson of the World Commission on Environment and Development, among other things), for her support. Norway positively responded to this initiative.

The Polish Interim Committee – in close collaboration with representatives of the Norwegian government and with the assistance of Coopers & Lybrand, a British consulting firm – drafted a comprehensive *Concept Paper* (Republic, 1991a) explaining the main points of the proposed solution. On June 11, 1991, at a special meeting attended by ambassadors of the Paris Club countries in Warsaw, Prime Minister Bielecki presented the concept paper. Thus he officially launched the initiative to coordinate prospective bilateral swaps within the framework of a new multilateral environmental fund that would add value to Poland's own efforts in areas of international and global concern. On the same day, Prime Minister Brundtland announced at a press conference that the Norwegian government would host the meeting of representatives of the Paris Club countries with representatives of the Polish government (as well as of key multilateral agencies) to be convened in Oslo on July 1, 1991. The aim of the conference was to discuss the Polish proposal for implementation of the Paris Club's April decision, and thus to pave the way for a series of bilateral negotiations (Republic, 1991b).

The Oslo conference revealed far-reaching support for the proposed model, the core financial mechanism of which is as follows. The EcoFund initially selects and evaluates projects eligible for funding (or co-financing) through the debt-for-environment swap. Next, the choice has to be approved by the participating Paris Club countries' representatives. The actual expenditures eligible for reimbursement from the EcoFund are then subtracted from a special escrow account with the Bank of International Settlements, where the Polish government services its external debt, or with an otherwise selected commercial bank. The amount invoiced – instead of being transferred to creditors' accounts – flows back to Poland to credit the EcoFund, which is therefore able to reimburse the successful project applicants.

From the very beginning, there have been three basic principles of the Polish debt-for-environment swap program: (1) additionality, (2) cost-effectiveness, and (3) "club approach".

Re (1). The projects selected are to be additional in the sense that, without the assistance of the swap programme, they would have either not proceeded at all or only proceeded at a substantially later date despite their international importance. In practice it is very difficult to assess additionality on a purely project basis since this presumes the existence of a clearly identified baseline program with a specific timetable for each individual project over the time horizon of two decades or so. For this reason the more relevant concept of additionality was defined in terms of total expenditure on environmental projects. The year 1991 – the most successful one until then – was declared as the benchmark for assessing future additionality.

Re (2). The cost-effectiveness approach – i.e. maximising the programme's objectives of addressing international environmental problems given available resources – was the core of the proposed financial mechanism. It assumed that projects would be selected and financed in such a way so as to achieve maximum results (in terms of pollution abated, habitats saved, etc.) per dollar spent. This implied, among other things, competitive bidding, which was also a condition to attract important multilateral co-financiers.

Re (3). The cost-effectiveness principle underpinned the Polish government's view on how the allocation of available money be best achieved. This should be on a "club" basis. By this, the government meant that each participating country would be able to indicate its preference between the different four areas of activity (i.e. – in short – acid-rain/transboundary pollution, Baltic, climate change, and biodiversity), and that individual preferences would then be added together to create an overall allocation of fund resources. Then, however, a detailed allocation between individual projects would be determined, taking into account technical considerations including cost-effectiveness criteria.

In order to emphasise the financial benefits for the creditor countries resulting from the swaps – in addition to the environmental ones – specific rules for procurement to be financed by the EcoFund were developed. On the one hand, for efficiency reasons, there should be no import purchases automatically tied to a creditor's commitment. On the other hand there is a need for creditor countries to perceive financial benefits, which should be proportional to commitments.

To resolve this dilemma, the Polish government offered to perform regular, periodic assessments of the geographical distribution of (import) contracts triggered by the EcoFund projects in the course of the competitive bidding process. Such a solution allows for the elimination of any "unfair" distribution of financial benefits accruing to contractors whose governments do not proportionally participate in the EcoFund initiative (Republic, 1992b).

It should be reiterated that there is an important difference between the approach outlined above, and a typical tied procurement scheme:

(a) the scale and the time profile of projects can be optimally adjusted to specific needs, and they do not have to follow a particular pattern of debt service to a given creditor;
(b) projects may typically require diversified import components, and the overall efficiency of an investment can be maximised only if various components are purchased from different contractors; this geographical differentiation would not have been possible under the tied procurement regime;
(c) EcoFund expenditures – as a result of the principle of seeking co-financiers for the project (including import components) – generate an *additional* demand for imports; this additional demand is greater than what could have

been sustained if the import were to be financed from the Polish budgetary sources.

From the economic theory point of view, the "club approach" proposed by the Polish government could also be interpreted as a way to overcome the inefficiency of free-riding, which is obviously preferred by non-participating countries. Some Paris Club members complained about constraining the procurement to the "EcoFund Club". Apparently they would prefer to take advantage of other creditors' swapped funds without contributing to the EcoFund.

The "club approach" was consistently adhered to by the Polish government not only because of its efficiency advantages (*vis à vis* tied procurement), but also because of the incentive it gave the non-participants to join. Occasional political pressures exerted by some creditors on Poland to withdraw from this approach, as well as the motivation informally acknowledged by latecomers confirm that this was an important design element. Initially it was only the Polish government who insisted on the "club approach". Later on, the participating countries had stakes in keeping this status as well.

After the Oslo conference the Polish Interim Committee worked out the institutional shape of the EcoFund, drafted its charter (Republic 1992a), and – with the help of consultants – wrote its detailed operations and procedures manuals (EcoFund, 1992a, 1992b). It was decided, among other things, that the EcoFund would be supervised by a Council with the collective (minority) representation of the participating creditor countries. On April 6, 1992, the Minister of Finance formally established the EcoFund as a foundation under Polish law and signed its charter. He also appointed the first six Polish members of the Council. The Minister of Environment appointed its Chairman. The Chief Executive Officer and the rest of the Board of Management were hired soon after that. In the Fall of 1992, the institution was ready to administer whatever resources would become available as a result of swap agreements. The Interim Committee was dissolved and its role was taken up by the EcoFund Council.

In 1991 and 1992 the government of Poland negotiated bilateral agreements with its creditors to practically implement the decision of the Paris Club. As the first contributor the American government confirmed that the EcoFund would receive 10 percent ($367 million) of Poland's indebtedness to the United States (in fact the bilateral agreement referred to a slightly higher American contribution). Also the French government agreed to swap its 10 percent, although it indicated that perhaps not the full amount of the swap would be channelled through the EcoFund. Agreements with other creditor countries were signed soon. However, apart from the Unites States and France, no other government took the advantage of the 10 percent option at that time. Some governments (e.g. Austria and Switzerland) explicitly confirmed the swap as an option for the future. Others merely referred to the Paris Club agreement, which leaves this open. In most cases, domestic budgetary problems were identified by creditor countries as the reason for no commitments. The following table summarizes creditors' contributions to the EcoFund.

Poland's indebtedness to the Paris Club countries

Rank		Debt (as of April 1, 1991)		Debt-for-Environment Swap	
		Million $	%	Year	Million $
3.	Austria	3719	11	-	-
13.	Belgium	336	1	-	-
5.	Brazil	3403	10	-	-
6.	Canada	2899	9	-	-
15.	Denmark	243	<1	-	-
16.	Finland	143	<1	-	-
2.	France	5171	15	1992	51.7
1.	Germany	6000	18	-	-
7.	Great Britain	2762	8	-	-
8.	Italy	1647	5	1998	32.6
9.	Japan	1276	4	-	-
11.	Netherlands	662	2	-	-
14.	Norway*	322	1	1997	0.1
17.	Spain	96	<1	-	-
10.	Sweden	613	2	1997	6.6
12.	Switzerland	528	2	1993	52.8

Rank		Debt (as of April 1, 1991)		Debt-for-Environment Swap	
		Million $	%	Year	Million $
4.	United States	3538	11	1991	367.0
	Total	33358	100	X	510.8

*Having been satisfied with the EcoFund's operations, Norway increased its contribution to 10% of debt due after 1998.

In 1993 the Swiss government signed an additional agreement to swap the full 10 percent (CHF 78 million, i.e. $52.8 million) of the debt. The French government decided finally to commit only 1 percent (FRF 280 million, i.e. $51.7 million) to the EcoFund, saving the rest for equity swaps. The next commitments were finalized as late as 1997 and 1998. In 1997 the Norwegian government donated NOK 1 million ($130,000) to the EcoFund, a gesture that was considered a "test run" of how the organization performs. Simultaneously the Swedish government committed SEK 50 million ($6.6 million), which corresponds to almost 2 percent of the Polish indebtedness to Sweden (the percentage of the Swedish commitment was related to the 1997 indebtedness rather than the original one, as Poland repaid a part of the debt due before this swap took effect). In 1998 the Italian government agreed to swap ITL 56 billion ($ 32.6 million), i.e. 2 percent of the original debt.

Germany was Poland's largest creditor. Consequently the German federal authorities – such as the government and the Bundestag – and environmental NGOs were an important target group of Polish lobbying efforts. At the same time they proved to be most reluctant to yield to such efforts. At the Oslo conference, the German government representative clearly stated the preference for purely bilateral assistance focused on the highly contaminated region of Silesia. Later on, German officials rejected the idea of a swap repeatedly on the grounds that, in 1989, Germany forgave a substantial part of the so-called "Jumbo Credit" (an export-promoting financial arrangement guaranteed by its government in the 1970s) and established a special Polish-German Foundation to administer the rest of the amount. The Foundation financed a spectrum of activities ranging from fostering youth exchange, to preserving German cultural heritage in Poland, to environmental protection.

In response, the Polish government indicated that the "Jumbo Credit" rescheduling deal was struck on the very day the Berlin Wall fell and it was a part of a broader agreement to compensate Poland for the expected loss of contracts implied by the disappearance of GDR. Besides, some other creditors had also reduced the Polish official indebtedness prior to the Paris Club decision in 1991; the latter thus reflected a compromise between what all the creditors considered justified given their respective histories with Polish debt agreements. Both the 50%- and 10%-clauses were understood by the Paris Club as independent of whatever had been bilaterally negotiated earlier.

Before the Polish response was publicized by political channels and also informally in professional fora, the "Jumbo Credit" argument was so popular that it influenced the overall German approach to the EcoFund. Even German environmental NGOs were once convinced that the federal government had already made a debt-for-environment swap as envisaged in the Paris Club decision. This conviction was shared by business community as well. The negative approach to the EcoFund proved fairly stable despite changing emphases. Later on, German authorities stressed the severe budgetary consequences of any future swap and explained its financial impossibility.

Virtually all the other Paris Club countries revealed interest in the EcoFund. In the case of the Brazilian government it was just an indication of interest of another heavily indebted economy contemplating the advantages of a debt-for-environment swap in solving its own predicament. Given the agendas of the official and business missions arriving in the Polish Ministry of Environment in 1991–1992, the rest of the group seriously considered some sort of a swap.

Internal political developments in Poland did not help nurturing foreign interests in the EcoFund at that time. The government that negotiated the Paris Club decision collapsed in December 1991. The next government survived until mid-1992 only. The term of its successor was a bit longer, but the 1993 witnessed another cabinet change. Between 1991 and 1993 the Ministry of Environment was led by four different ministers and one temporary administrator (who substituted a Minister who quit before the end of his term). This no doubt created uncertainty regarding the seriousness of the Polish commitment to run the EcoFund according to the principles outlined in 1991. Perhaps some countries might have doubted if the Polish government would maintain its declared preference to seek debt-for-environment swaps only. At least one European government was close to committing to the EcoFund in 1992, but it was discouraged by the political uncertainty.

All major Polish environmental NGOs declared their support for the EcoFund. They assisted the government by lobbying their fellow groups among the Paris Club countries. As a result, many international networks, such as the Worldwide Fund for Nature (WWF) and Friends of the Earth, gave their blessing and promised to encourage their national member organizations to push for the Polish swap. Indeed in several countries there were seminars, press releases and cabinet briefings arranged by local NGOs. For instance, the Swedish-Polish Environmental Organization (SPM) organized and financed two meetings with officials and Riksdag members devoted to presenting the idea of the EcoFund. Even in Germany, where the initial response by NGOs was reluctant, environmentalists did exert some pressure by publicizing EcoFund activities (Ribbe, 1994).

Some NGOs tried to reach local business communities to discuss the "club approach". In most cases business leaders did not see any merit in departing from the tied procurement philosophy applied in the conventional bilateral assistance schemes they were used to. Businessmen sensed that a debt-for-environment swap with Poland would divert resources allocated to bilateral assistance. It was clear from their reactions that they preferred their governments not to join the EcoFund. The only motivation for revealing any interest in the swap was linked to the possibility of losing potential contracts to firms whose governments did join the EcoFund. The favourable business press coverage that the EcoFund seems to have enjoyed (e.g. Rohwedder, 1993; Bajsarowicz and Zylicz, 1993) was not a factor that changed this general attitude.

Several countries – e.g. Sweden – indicated that if a swap was made, the money would be diverted from the existing bilateral assistance programs. While this was precisely what the Polish government advocated for, the trade-off was considered unacceptable for some stakeholders in the creditor countries. For some countries, multilateralism was thus an important obstacle to participating in the EcoFund. This factor could be relevant especially for countries with extensive bilateral assistance programs. One can speculate that vested interests linked to the tied procurement prevented e.g. Denmark (the largest bilateral donor) from switching to the "club approach".

Another interesting feature of the debt-for-environment swap diplomacy was issue linkage. Bilateral consultations took place at various fora, so that very different issues were discussed at the same time – perhaps informally. There was, however, at least one instance of a serious linkage; a creditor wished to connect the contribution to the EcoFund with a deal favouring its truck manufacturer. The Polish government did not accept the offer.

The EcoFund completed its mission in 2010 and the foundation was formally liquidated in 2012. Its financial contribution to Poland's environmental protection was not high, although its impact has nevertheless been considerable and lasting. First of all, it paved the way for establishing a domestic institution to address international priorities. More importantly, it promoted the concept of cost-effectiveness, which was then taken over by other Polish financial institutions.

The prospects for replicating the Polish EcoFund pattern are slim. The (limited) success of this debt-for-environment swap required a strong commitment from the debtor and a fair degree of trust on behalf of participating creditors. Neither of these factors is easily replicable. Encouraged by the Polish example, the Swiss government established a similar debt-for-environment swap facility in Bulgaria, but no other creditors were willing to participate. Both the lack of strong pressure from the Bulgarian government and the free-riding of other creditors could be factors that prevented this initiative from developing into a larger-scale arrangement. This illustrates just how difficult it is to establish a cost-effective multilateral mechanism to provide an international public good.

4.2.3 Applied game theory

International environmental agreements have been discussed in terms of game theory concepts at least since the 1980s (e.g. Mäler, 1989; see Mickwitz, 1998 for a review of how actual policies in Europe followed theoretical – including game theory – principles). Debt-for-nature swaps were subject of many analyses, but there were only few attempts to interpret them as games. Chambers *et al.* (1996) developed a game theoretic model of a debt-for-nature swap. They portrayed it as a second best solution to the problem of global deforestation in the absence of a supranational entity to enforce the global optimum. The key players in that game were environmental NGOs who let their members pay for saving tropical forests. In contrast, in the Polish EcoFund case, the initiative is at the debtor side, and the swaps were carried out by creditor governments rather than NGOs (even though the latter might have interacted with their governments and with each other).

The Polish debt-for-environment swap initiative can be interpreted as a sequence of two games. The first game relates to whether equity or environment is to be swapped. This game was played by the Polish government with its creditors in 1991–1992. The second one relates to the participation of a given creditor in the debt-for-environment swap facility that was already established. That one has been played (with non-participants) for many years. While a more realistic representation of both games would require a much more complex analytical framework (Folmer *et al.*, 1993), quantification problems would render this approach impossible to validate in this case. Therefore this paper applies a simplified method that separates the two stages of the swap development.

In the first stage, Poland played with all creditors the same game separately. The games took place during bilateral debt rescheduling negotiations. The Paris Club decision mandated 50 percent reductions, but technical details were left to bilateral agreements. The Polish government took this opportunity to approach each creditor individually and suggested taking advantage of the additional 10 percent clause. Three main options were available for either party:

- no swap (NS);
- debt-for-equity swap (EQ); and
- debt-for-environment swap (EN).

One can contemplate a 3x3 payoff matrix for the parties involved in such a game. The payoffs are understood as perceived by swap decision makers – mainly finance ministers, governments and parliaments. That is, they may not necessarily reflect full costs and benefits of various strategies. Both creditor and debtor are considered homogeneous entities. Given the logic of political process, a homogeneous creditor may therefore tend to focus on financial benefits rather than environmental ones.

Hypothetical payoffs in a creditor-debtor game

		Creditor		
		NS	**EQ**	**EN**
Debtor	NS	0,0	0,0	0,0
	EQ	0,0	*a,b*	*c,d*
	EN	0,0	0,0	*e,f*

It is assumed that a swap takes place if parties agree on the debt-for-environment swap (EN-EN), or if the debtor chooses the debt-for-equity swap (EQ-EQ or EQ-EN). In the latter case, the swap takes place even if there is a mismatch between the preferences of both parties. For if the debtor accepts a debt-for-equity swap, then a debt-for-environment swap can be carried out as well, with the creditor not only providing environmental benefits, but also retaining ownership of a facility (project) producing such benefits. In other words, swaps take place unless a party chooses the no-swap option, or the debtor insists on a debt-for-environment while the creditor accepts an equity swap only.

The payoffs are understood as incremental with respect to the no-swap option. Apart from the trivial outcome (NS-NS), the game has a Nash equilibrium (EN-EN) with the payoffs (e,f) as long as $e \geq c$, and $e,f \geq 0$. If, in addition, $a<0$, then (EN-EN) is the only non-trivial equilibrium. If $a,b \geq 0$, and $b \geq d$, then (EQ-EQ) is another Nash equilibrium with the payoffs (a,b).

A solvent debtor may perceive the incremental benefits from a debt-for-equity swap as non positive, since a creditor is presumably interested in assets that would have found a buyer anyway. Moreover, unless the creditor buys assets in a competitive bidding procedure (a rather unlikely arrangement under the swap), the purchase price can be lower than the market value. Hence, one can assume that $a<0$.

Now, assuming that a creditor is "environmentally biased", then an equity swap can include an asset whose market value does not capture all the benefits provided through its operations. Hence, $c>a$ seems a justified conjecture (although it does not necessarily follow from this that $c>0$). Moreover, a debt-for-environment swap provides a debtor with maximum and positive benefits, as it does not require any ownership transfers. Hence $e \geq c$, and $e>0$.

The analysis of costs and benefits from a creditor's point of view looks certainly different. It can be argued that $b \geq d \geq f$ owing to the fact that the less constrained the creditor is in spending decisions, the more likely it is that a profitable spending mix will be selected. Nevertheless, even under the most constrained regime – i.e. a debt-for-environment swap without any transfer of the debtor's assets – the creditor may receive some net benefits. This justifies the conjecture that $f \geq 0$.

In assessing their benefits, creditors take into account both environmental and financial gains from a swap net of expenditures. Environmental benefits depend on a number of circumstances as explained in 4.2.1. In particular, *ceteris paribus*, they are higher for a downwind/downstream party than for an upwind/upstream one. Financial benefits depend on how successful in triggering additional contracts is the swap program. This additionality, in turn, is caused by a so-called leverage the swap can achieve. The leverage is observed whenever an expenditure from a given source – here from the creditor's budget – activates someone else's spending, e.g. spending from the debtor's budget or from the private sector. Other things being equal, financial benefits are higher for a country whose firms were not very successful in the debtor's market earlier. The swap may open up the market for them. On the contrary, the country whose presence in the local market has been firmly established will not gain as much from the stimulus provided by the swap.

To sum up, the model predicts a debt-for-environment swap as a Nash equilibrium in a game between a solvent and environmentally conscious debtor and a creditor without a strong, established presence in the debtor market. At the same time the model anticipates that no swap will take place if the creditor is successful in the local market anyway. Finally, the model explains why a debtor may be better off without a swap rather than with a precedent-setting debt-for-equity arrangement which makes other creditors expect a departure from the debt-for-environment preference.

In the second stage, all the creditors recognized the firm preference of the Polish government to negotiate debt-for-environment swaps rather than alternative arrangements left open by the Paris Club 1991 decision. Once the Polish strategy became fully disclosed and known, the only remaining game was that to be played by the creditors. Given the multilateral design of the EcoFund, the outcome of participation of any given creditor depended on the total number of participants (and their commitments). Therefore, our analysis of the second stage draws from the analysis of self-enforcing agreements as developed by Barrett (1994).

Each member of the Paris Club takes a decision whether to participate in the EcoFund or not. To this end, payoffs from both participation and non-participation need to be assessed. It can be assumed that while, in first case, the payoff decreases with the number of participants, in the second case it increases.

For any given creditor, the payoff is the difference between the benefit from the swap and its cost. While the latter is independent of the number of participants, the former is not. For a participant the cost is the amount of debt forgiveness, and for a non-participant the cost is zero. Both participants and non-participants derive environmental benefits from swaps. The level of such benefits is perceived as low by creditors. Financial benefits are by and large perceived by them as essential for the participation decision.

The benefits from participation include, most importantly, the value of contracts obtained through the EcoFund. Thanks to the co-financing requirement, each dollar spent by the EcoFund generates three to four dollars from other sources. Hence creditors can count on this "leverage" mechanism providing more revenues than the funds swapped. As long as the EcoFund is a small facility, demand for its support is much larger than supply, and project applications are usually accepted only when they demonstrate significant co-financing from other sources. Creditors may be afraid of weakening the "leverage" effect if the supply of EcoFund resources is increased. This would lead to lower financial benefits from participation.

Non-participants face the barrier of the "club approach". Only those who contribute to the EcoFund are eligible for contracts supported by the debt-for-environment swap. There is, however, an efficiency-motivated provision that softens this barrier. Up to 10 percent of the EcoFund's spending on projects can go to non-participating contractors who supply some complementary equipment or services at a price lower than firms from participating countries. This provision lets non-participants enjoy benefits from the debt-for-environment swaps in a free-riding manner. Moreover, the larger the EcoFund's budget, the higher the benefits for the (fewer) remaining free-riders. Hence benefits from non-participation increase with the number of participants.

The argument just outlined warrants the following payoff functions for participants and non-participants, R_p and R_n, respectively:

- $R_p(s) = g - hs$ (for participants)
- $R_n(s) = j + ks$ (for non-participants)

where s is the number of countries participating in the swap, and $g, h, j, k > 0$ are constants. The notation implies that countries are identical with respect to their parameters of the functions R_p and R_n. We make this assumption just to obtain a preliminary insight into the mechanism of joining the swap.

A creditor country has an incentive to join the swap if the benefits from its participation outweigh the benefits from non-participation, i.e. $R_p(s) > R_n(s-1)$. Solving this inequality for s, we obtain $s < (g-j-k)/(k+h)$. Let us define $s^* = [(g-j-k)/(k+h)]$, where $[x]$ denotes the *entier*, the largest integer not greater than x. Then s^* is the largest number of countries that the EcoFund can sustain.

If $g \geq j+k$ then $s^* \geq 0$. Otherwise not a single country would have joined the swap. Apparently $s^* \geq 6$, since six countries have joined the EcoFund as long-term members of the "Club". If $s^* \geq 17$, then the entire Paris Club would have benefitted from the swap.

A more realistic analysis requires that creditors are perceived as non-identical. One way to depart from the assumption made earlier is to define s as the amount of money contributed to the EcoFund rather than the number of countries. The equilibrium condition for s^* would then need to be reinterpreted as a condition for increasing a creditor's contribution to the EcoFund. This approach could shed some light on creditors' decisions to contribute less than the 10 percent of the debt. (Three countries out of the participating six indeed chose such an option.) Further improvements of the model require lifting the assumption that g, h, j, and k are the same for all countries. As in stage one, we can observe that, *ceteris paribus*, being a downwind or downstream country results in higher benefits from participation while having a strong established position in the debtor country's market results in higher benefits from non-participation. However, the lack of quantitative estimates of benefits would make such an exercise hardly revealing.

The formal model analysed here seems to be consistent with the Polish experience. The EcoFund history confirms that environmental benefits provided through the swaps were of secondary importance for the creditors. The absence of Germany and Denmark – Poland's close neighbours, who share the same water resources – is remarkable, but there might have been very peculiar reasons for those countries not to contribute. Much more telling evidence is the total *désintérressement* of participants with respect to the rules of allocating money between the four areas of EcoFund activity. It was envisaged in the original design that the creditors would state their environmental preferences by targeting their contributions at specific areas and determining in this way the money available for each of these. It turned out, however, that none of the creditors has ever taken advantage of this provision, thus demonstrating indifference towards specific environmental benefits.

The model also suggests that there may be a maximum number of participants (lower than seventeen) that the EcoFund can sustain due to free riding by non-participants. The provision that allows non-participants to bid for contracts up to 10 percent of the EcoFund spending was conceived as an efficiency-enhancing instrument. Nevertheless, to some degree it weakens incentives to join the swap. As in the case of self-enforcing agreements, the larger the benefits available to free riders, the lower the number of participants. One practical recommendation that could have been derived from this argument was to confine the procurement strictly to the "Club" by eliminating even the 10 percent margin kept for efficiency reasons. The growing diversity within the "Club" made this margin less important.

4.2.4 Lessons learnt from the Polish debt-for-environment swap

The global experience with debt-for-environment swaps is severely limited, as almost 99 percent of debt conversion transactions have taken the form of debt-for-equity swaps. Creditors prefer the latter, since they are more similar to bankruptcy procedures, which solve indebtedness problems at the national level in most jurisdictions. Nevertheless, it has been argued that the former are potentially very useful as a means to address global environmental problems in countries whose economic performance and environmental awareness are unlikely to guarantee due appreciation of these problems.

Debt-for-environment swaps are uniquely suited to introduce global environmental priorities into consideration in low and middle income countries. They can help articulate both international and national environmental concerns, and – by promoting the issue at national government fora – to seek available funds to address both. By establishing professionally managed facilities to administer swapped funds, they can also assist in local capacity building, which is a prerequisite for carrying out environmental protection on a sustainable basis.

In the Polish debt-for-environment swap initiative of 1991, several issues were addressed in a way that made the EcoFund a model for a facility supporting projects of global importance:

- emphasis on creating a domestic professional institution, as opposed to *ad hoc* commissions, hired consultants who come and go, etc.;
- multilateralism as opposed to a series of difficult-to-coordinate bilateral deals;
- serving creditors' financial interests by *ex post* ensuring fair geographical distribution of contracts rather than inflating costs by *ex ante* tied procurement.

Praised by environmental NGOs and international financial institutions, the Polish swap has nevertheless enjoyed very limited support from Paris Club governments. The game theoretic model discussed in 4.2.3 explains why certain creditors may find it profitable not to join, especially when others do and thus let the EcoFund operate. Multilateralism, one of the key features making the swap a cost-effective mechanism, is responsible for the lack of support by many creditors at the same time.

Any sincere appraisal of a debt-for-environment swap must take into account its environmental benefits. The political reality is that decisions on whether to enter such a swap agreement are taken by finance ministers rather than bodies representing a fully balanced array of its effects. As a result, financial benefits tend to be the major factor guiding creditor countries' policies in this respect. Benefits of that kind do exist, but their distribution is not restricted to participants which provides non-participants with wrong incentives.

Looking from the debtor's side, one sees a mixture of environmental and budgetary consequences of a swap. Both are important and both were taken into account by the 1991 government – the greenest of Poland's governments after 1989. One can speculate what would have happened if a swap opportunity was debated by one of the other Polish governments that ruled after the 1989. While a swap would have been accepted, it is almost certain that this would have been a kind of a conventional debt-for-equity arrangement advocated by the strongest interest group at any particular time (privatization? highways? agriculture?).

4.3 Environmental aspects of European non-environmental policies

Issues such as e.g. cohesion and regionalism are of paramount importance because of political reasons. They also have important implications for environmental protection as organized in the European Union. This is a group of countries which do not have a central government, and whose budget is around 1% of the regional GDP. From the point of view of environmental protection, it is an interesting case as it combines certain characteristics of self-enforcing agreements, and centralised policies.

The following analysis demonstrates traps that come with policies aimed at introducing uniform standards for large political units (including nation states as well). The noise emitted from lawn mowers is regulated in the European Union by five directives: 84/538/EEC, 85/409/EEC, 87/252/EEC, 88/180/EEC, and 88/181/EEC. Its permissible level is regulated uniformly.

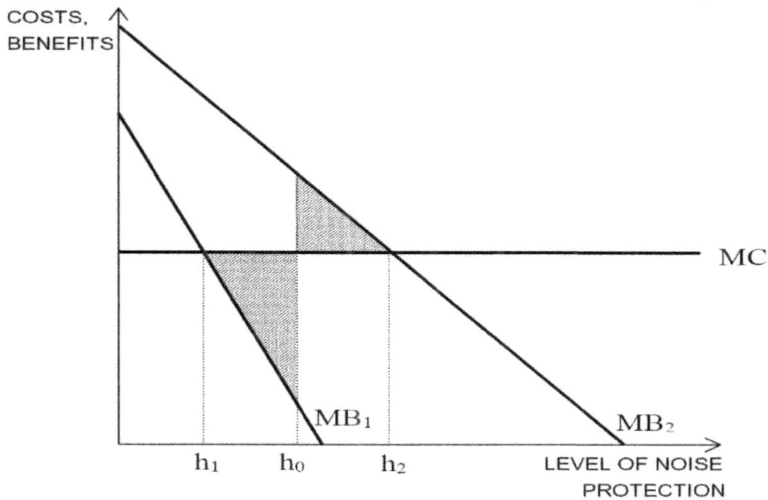

It can be assumed that the Marginal Cost (MC) of increasing the level of noise protection is constant, while its Marginal Benefit (MB) is declining. For simplicity, let us also assume that there are two countries: no. 1 where the people are not very upset by the noise, and no. 2 where the people are willing to pay more in order to reduce the noise. The lower number h_1 is the level of noise protection justified in the first country, and the higher number h_2 is the level of noise protection justified in the second country. Seeking a "compromise", an average h_0 of the two numbers, can be suggested as a uniform standard. This standard, however, lowers the welfare in either country, as illustrated by the shaded area: the first country has to pay a cost which is higher than its benefit, and the second one loses benefits which are higher than the cost saved.

If a uniform standard is welfare-decreasing, then one may ask why countries agree to adopt it in the first place. The analysis above does not reflect non-environmental benefits from having a uniform standard. These may be high, as they allow for better economic integration. A lawn mower that was certified in one country becomes legal in another one too. This saves on transaction costs and makes the life of manufacturers easier. Hence studying environmental protection at the European level must take into account other aspects of regional cooperation.

Since the very beginning, "cohesion", i.e. making regions similar to each other, has been an objective of the European Union. This similarity, however, needs to be interpreted adequately. It is not about making regions look similarly. It is about providing people with equal opportunities to support their aspirations without impairing diversity, which has been an important asset of the Union.

Translated into policy instruments, "cohesion" was understood as transferring wealth from regions that record GDP per capita higher than average to those that are below average. It was also meant as a way to let those that are below average approximate those above permanently, i.e. to let them build a potential for sustaining wealth rather than relying on transfers. Hence "cohesion" policies aimed to increase employment and innovativeness in depressed regions. They aimed to establish broadly understood convergence both at the national and sub-national levels.

One clarification that needs to be made is about the level of disaggregation. European nation states range from very small units, such as Malta, to large economies, such as Germany. It is thus obvious that some disaggregation of big geographical areas is necessary. A compromise between size evenness and data availability has to be struck. This is accomplished by using so-called NUTS1, NUTS2 and NUTS3 networks. NUTS1 lists 98 units (large countries are represented by several units). NUTS2 lists 272 units (often smaller than German federal states). NUTS3 lists 1325 units. Most of the "cohesion" activities are aimed at the NUTS2 level. Only some specific programmes are run at the NUTS3 level.

Another clarification refers to how to measure "cohesion". There are basically two concepts of convergence: "beta" and "sigma". The former refers to a faster change rate observed for units below the average than for those that are above.

The latter refers to lowering dispersion between the units. It can be proved that "sigma" convergence implies "beta" convergence, but not *vice versa.*

European "cohesion" policies have proved to be little effective at sub-national levels. Statistical research (Paas and Schlitte, 2007) demonstrates that "sigma" convergence is observed at the national level. However, both at the NUTS2 and at the NUTS3 level, GDP per capita "sigma" diverged. "Beta" convergence was observed internally in a few countries only. In most countries GDP per capita "beta" diverged.

The question arises whether income convergence is correlated with environmental protection, as many analysts believe. The problem is linked to the so-called Environmental Kuznets Curve (EKC), and it was found that the relationship between GDP per capita and environmental disruption resembles an inverted "U". In other words, when GDP per capita is low, its increase leads to higher environmental disruption, but when it is sufficiently high, its further increase leads to lower environmental disruption.

Environmental Kuznets Curve

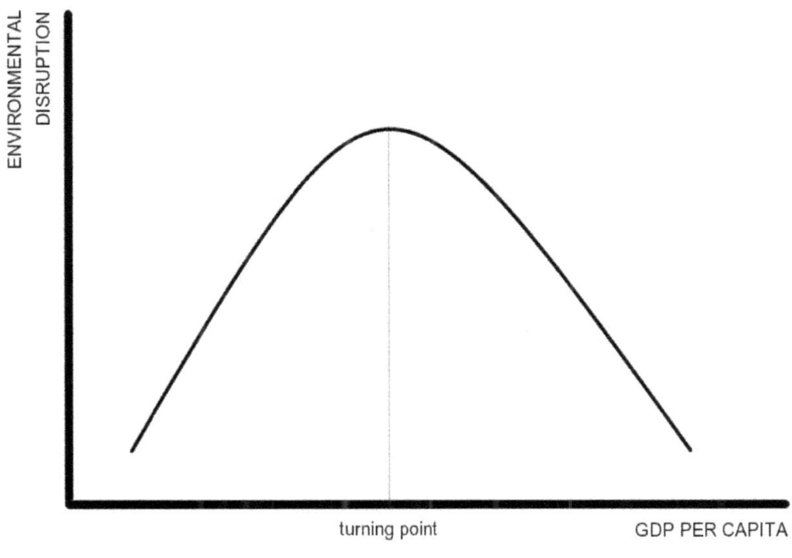

EKC has been empirically verified for dozens of data sets. The pattern is not as regular as a standard picture illustrates, but indeed very often one sees environmental disruption (measured by deforestation, pollution, or whatever) first growing, then stabilising, and eventually declining. The most interesting element of these analyses is the identification of the turning point. Sometimes empirical observations are located in the left-hand side of the picture, so only the curvature of a fitted graph suggests that there may be some turning point. This is the case of carbon dioxide emission, for which the turning point is calculated well above observed GDP per capita even in rich countries (Jobert *et al.*, 2010). For other aspects of environmental degradation, observations exist both to the left and to the right of hypothetical turning points. In many applications – as e.g. for instance in the case of sulphur dioxide – the turning point is found around $5000 in 2000 prices (Frankel, 2009).

Even though EKC can be detected in some cases (Johansson and Kristrom, 2007), a great variety of additional factors may determine an actual relationship between GDP per capita and environmental disruption (Kukla-Gryz, 2009). Nevertheless, for many aspects of environmental degradation there may be a broad tendency to have it reduced once GDP per capita grows after certain threshold that European countries have already passed. Thus European "cohesion" policy can be seen as environmentally friendly. There are two exceptions though. One is biodiversity and another is climate protection. To the extent that biodiversity is reduced by switching from traditional to modern agriculture, getting richer does not solve the problem. Despite the fact that, in some economies, energy use has been decoupled from GDP growth, as a rule wealthier countries have much higher per capita energy consumption than others. The European Union is aware of these trends and amends its "cohesion" policy by programmes dedicated to biodiversity conservation and fossil fuel elimination.

In order to slow down biodiversity loss, the European Union launched its Natura 2000 programme, which is based on two European directives: 79/409/EEC on the protection of birds, and 92/43/EEC on the conservation of habitats. In 2011, it protected around 18% of land in the EU countries, and covered the most valuable terrestrial habitats. Natura 2000 is also being enlarged in the marine off-shore environment.

Less wealthy European countries sometimes treat the Natura 2000 network as a mismatch between environmental preferences of the "rich" and the "poor", whose aspirations are linked to the consumption of material goods. Given the fact that industrialization in most of Poland started later than in Western Europe, natural habitats are less scarce here than elsewhere. On the other hand, material welfare is lower than in Western Europe. In combination, these two circumstances result in lower economic value placed on the natural environment in Poland than in wealthier societies. Nevertheless, Poland's implementation of the Natura 2000 network has been quite effective.

In 2013 the total number of Natura 2000 sites was 967, and their area reached 9.4 million hectares, i.e. 30% of the country. Initial implementation of the network was slow because it was perceived as a foreign idea imposing wealthy country preferences on a developing country whose priorities are obviously different. A frequent concern was that obtaining Natura 2000 status precludes economic development of a site.

This opinion was so pervasive that, for several years, the government hesitated to establish Natura 2000 sites beyond the absolute obvious. As a result, environmental NGOs produced their own "shadow list" and lobbied the European Commission to reject the official list. Meanwhile, public opinion in Poland had an opportunity to gain an appreciation for the system. Some engineers advocated for an extremely inefficient highway in north-eastern Poland. The highway was supposed to cross the Rospuda Valley, an especially valuable peat bog that was an officially designated Natura 2000 site. The European Commission ruled the project illegal. Closer scrutiny then also revealed its drastic economic inefficiency. The case seems to have demonstrated that the system not only protects biodiversity, but also human welfare. The government took the "shadow list" and accepted the system in its entirety.

Despite the fact that Natura 2000 has become a part of Polish government policy, some tension between high-income and middle-income country preferences persists. Polish politicians feel that legal requirements impose constraints on the sites' economic development that are not justified by local preferences. To the extent that uniform policies or uniform regulations are not economically justified, this may be true. On the other hand, Natura 2000 regulations do not introduce new protection regimes; they rely on protection regimes which exist in a given country's legislation – in the Polish case national park, landscape park, or nature reserve. Consequently, specific development constraints can be adjusted by flexibly choosing a specific protection regime.

The integrity of a Natura 2000 site can be compromised under very restrictive conditions. These include an important social objective, and the lack of an alternative solution. A typical welfare-decreasing project professes to serve some important social purpose, whereas in reality it does not. It also claims to be the only alternative available, even though alternative solutions were not sought. Thus Natura 2000 introduces a rigor to be explicit about the social objectives a given development project serves, as well as to verify whether these social objectives can otherwise be served.

The Natura 2000 program helps to articulate local priorities. If a middle-income country deems that high-income countries are imposing alien priorities, then it must carefully assess its own development preferences and opportunities in order to explain why a specific protection regime is inadequate. As the program does not prescribe fixed solutions, it is flexible, and welcomes rather than dampens the expression of local preferences.

In contrast, the European Union climate-related policies are not that flexible. In principle, ETS could play the role of a flexibility mechanism, but its function is compromised by numerous administrative interventions aimed at "correcting" the trades. In addition, environmental NGOs occasionally table proposals on how to spend the revenues from carbon dioxide permit auctions, which undermine the credibility of the system.

The European Commission runs a number of programmes to support the development of renewable energy. In particular, renewable energy is allowed to be subsidized, despite the general principle of strict limitations for public support. This might be justified given external benefits of renewable energy, but it is not well coordinated with other policies, in particular with climate-related instruments. As a result, European producers of renewable energy installations lost their competitive advantage to firms from non-Annex I countries. To remedy the situation, the Commission looks for some solutions that may call for further interventions, and so on.

An interesting case for seeking synergies between various policies was the Lisbon Strategy, which was adopted in 2000 as a strategy to "make Europe the most competitive and the most dynamic knowledge-based economic region of the world by the year 2010". Relevant policy documents were amended in 2001 by adding that competitiveness and dynamism should be consistent with sustainability. The authors of the strategy were of the opinion that economic aspirations could be reconciled with environmental criteria. Consequently, the Lisbon Strategy had a very solid environmental pillar. It consisted of four objectives:

1. Climate change; slowing down fossil fuel consumption;
2. Conservation of natural resources;
3. Mitigating transport pressure; and
4. Improving public health.

Its mid-term assessment revealed that the environmental pillar was perhaps established too optimistically. The four objectives turned out to be more difficult to implement than expected. The first one attracted most attention. Indeed the European Union was acclaimed the world leader in area of the Kyoto Protocol, but the results proved to be problematic. As explained in 3.2, global carbon emissions kept growing, and instruments unilaterally applied succeeded in reducing production-based carbon footprints, but not consumption-based ones. Energy efficiency improved, which did not lead to higher competitiveness. Conservation of natural resources had two aspects. With respect to living resources no progress was achieved; biodiversity loss has continued. However, the effectiveness of waste management improved. Mitigating transport pressure was a total failure. Further improving public health did not progress either, although this was caused by the fact that public health was in a fairly good shape at the outset already.

In 2004 a committee was appointed to review the Lisbon Strategy and recommend its revision for the second half of its period, i.e. for 2005–2010 (*Facing*, 2004). The revised Lisbon Strategy was "freed" from most of its environmental ambitions. The entire environmental pillar was reduced to Guideline no. 14, out of the total of 23, which read:

To encourage the sustainable use of resources and strengthen the synergies between environmental protection and growth.

The revised Lisbon Strategy was dominated by the search for ways to encourage competitiveness, and environmental issues were accounted for only to the extent they could be expected to deliver some purely economic objectives (like GDP or employment). More specifically, Guideline no. 14 recommended:

- Internalising external costs;
- Increasing energy efficiency; and
- Support for environmentally-friendly technologies (ETAP).

The first recommendation is certainly sound, but it is difficult to implement, as recent decades have demonstrated. Also, the second one is consistent with common sense, but – as the rebound effect (see 3.2.4) warns – it may send the wrong signals because while economic agents are convinced that they move in the right direction, environmental protection is not furthered. The most dangerous recommendation is the third one. As the ETAP (Environmental Technology Action Plan launched by the European Commission in 2006) demonstrates, any such programme can be captured by lobbyists who see this as an opportunity to promote their technology by convincing administrators to designate it as "the best" solution to a given environmental problem. Even though a given technology could be the best solution at any given time, it does not necessarily retain this status in the future. It is quite likely that under new circumstances – such as changing preferences and prices – another technology will turn out to be better. Thus, administrators could serve their constituencies more adequately by looking at environmental problems rather than specific technologies. However, if the policy is dominated by business considerations, as the revised Lisbon Strategy was, then the only *raison d'être* for doing something environmental is a commercial interest. Environmental concerns were overshadowed by economic ones.

According to the mid-term report (*Facing*, 2004), the "European Union and its Member States have clearly themselves contributed to slow progress by failing to act on much of the Lisbon strategy with sufficient urgency. This disappointing delivery is due to an overloaded agenda, poor coordination and conflicting priorities." The report anticipated that, once freed from an environmental ballast, the strategy would achieve its purely economic objectives. In 2010, upon its completion, it was anyway declared a failure by many European political leaders, though there was no agreement as to its causes. But one thing was certain: it has not failed because of "overloading" with environmental objectives. Hence, as environmentalists believe, it may be possible to combine environmental ambition with economic performance.

4.4 Trade and environment

This was a fashionable topic in the early 1990s. In 1993 the Swedish government adopted a special report, and in 1994 *Ecological Economics* published a Special Issue (Ekins *et al.*, 1994). The Swedish government report (Andersson *et al.*, 1995) concluded that trade is neither good nor bad for the environment; as any instrument, it is neutral. If applied skilfully, it improves things, but if applied inadequately, it can worsen the predicament. The fashion could have been influenced by a scandal caused by a 1991 internal memo by Larry Summers (the chief economist of the World Bank), which included the following famous wording: "the economic logic behind dumping a load of toxic waste in the lowest wage country is impeccable and we should face up to that". The memo was leaked to the press and ruined Summers' career in the Bank.

The direction of causality for dirty industries to be found in countries with lax environmental regulation is not clear. One may argue that dirty industries go where environmental regulation is not strict. But it can be also argued that strict environmental regulations are not adopted in countries where dirty industries exist, since politicians would hesitate to struggle with their lobbyists. Hence the trade-environment nexus is far from obvious.

Economic theory does not preclude that the trade-environment relationship may go either way. So-called eco-dumping (i.e. production subject to poor environmental requirements) has ambiguous welfare effects for the country which decides to adopt a low environmental standard (Althammer and Buchholz, 1995). Adopting an environmental treaty may have influence on a country's competitiveness. Therefore it can be analysed in order to see whether the treaty implies an economic success. The stability of international environmental agreements is analysed using sophisticated economic models, in particular using their trade repercussions (Diamantoudi and Sartzetakis, 2006). Likewise international trade can be beneficial, just as it can be detrimental to environmental protection (Bhagwati and Srinivasar, 1996). In what follows, theoretical linkages are outlined and confronted with statistics.

4.4.1 Academic analyses

Economic theories linking environment and trade usually refer to Environmental Kuznets Curves. Unless a country is very poor, income growth leads to better environmental protection. Given that considering countries at low income levels permanently is not a realistic and attractive option, most analyses refer to a world with economies wealthier than the "turning point" (see 4.3 above). Thus, it is assumed that growing wealth results in better environmental protection.

At least since the early 19th century, economic analyses demonstrated that trade is a prerequisite for growing affluence (Smith, 1776). David Ricardo (1817) provided the classical argument – contradicting earlier assumptions that exchange is always based on a fraud – that both parties to a transaction may benefit simultaneously. The argument refers to so-called comparative advantages. In the 20th century, a Swedish economist, Heckscher, and his student, Ohlin (1967), extended the Ricardian argument by linking comparative advantages to endowments in production factors – such as soil or hydropower – that can be country-specific. This theory explains how international trade leads to growing wealth.

The next step is to apply the standard EKC argument in order to conclude that international trade contributes to environmental protection. Thus, the sequence of implications is as follows:

TRADE → ECONOMIC SURPLUS → ENVIRONMENTAL PROTECTION

Critics point out that either of these implications may fail in real world circumstances.

International trade may be harmful for the poorer party. This is a so-called specialization trap that many developing countries fell into. An economy reduces autarky in order to specialize in what is profitable. It exports some goods and imports whatever else is needed. The trap may appear for two reasons. First, exporters from the poor country lack the capital necessary to flexibly switch to an alternative sector if the terms of trade (the relation between export and import prices) decline. Consequently, the decline in profitability does not imply lower export. On the contrary, in order to afford the same volume of import, the country has to export more, despite less attractive revenues. Second, if export relies on agriculture – as it often does – exporters tend to reduce biodiversity by switching to single crops that are more vulnerable to pest infestation. In order to survive, they have to import growing amounts of pesticides which ruin their bottom lines (not to mention local ecosystems). Irrespective of a specific route leading to a specialization trap, trade does not necessarily lead to growing income.

Now, let us look at the second implication. The Environmental Kuznets Curve cannot be taken for granted even if GDP per capita is higher than the "turning point". The standard argument assumes that greater average income translates into better environmental quality caused by higher supply of, and higher demand for, protection. While higher supply is linked to better technologies, higher demand results from a natural preference for better environmental quality once other needs are sufficiently satisfied. These mechanisms do not work, however, if greater wealth is linked to a more uneven distribution (in other words, if the rich get richer and the poor stay poor).

Those who become richer may not increase their demand for better environmental quality, as they can buy a better environment for themselves and their families in other ways. They do not have to demand higher quality tap water, since they can afford to buy bottled water. They do not have to demand better air quality, since they can afford to send their children to school abroad. They do not have to demand landscape protection, since they can afford to buy a nice piece of real estate and fence it off from unpleasant views. A similar mechanism prevents the supply of environmental protection from being enhanced. Unless the rich spend their money domestically, the local economy does not have to benefit from technology development.

There are also some non-income-based arguments (both negative and positive) that shed light on the trade-environment debate.

Negative arguments refer to environmental externalities similar to those mentioned earlier. Export specialization favours monoculture, which is likely to imply external costs in terms of lower biodiversity and higher contamination with chemical residues. Traditional autarkic agriculture may be less beneficial for farmers, but it is definitely more environment friendly. Another argument – not linked to agricultural externalities – is called the "race to the bottom". It refers to capital transfers rather than to trade in commodities, and envisages lowering environmental requirements so as to make an area more attractive for prospective investors. Given that freedom of trade is closely linked to the freedom of general mobility, the "race to the bottom" should be noted here, too.

Positive arguments may refer to technological spill-over effects. By being involved in international exchange, domestic producers are likely to apply more resource-efficient technologies. Another argument refers to the exposition to more environment friendly standards. They may not be fully justified by existing local preferences, but the preferences themselves are likely to evolve, resulting in more emphasis placed on environmental quality. Ultimately environmental protection may benefit from the local economy being exposed to international trade.

Theoretical arguments may go either way. Trade is either good or bad for the environment, and the theoretical conclusion is likely to depend on an analyst's *a priori* convictions, but the problem has been researched empirically as well. The following review is mostly based on the 1990–2004 data (Frankel, 2008).

International trade is present in all economies. The first issue to be checked is whether the trade-to-GDP ratio – a measure of the role international trade plays in a given economy – is correlated with environmental protection. There are two distinct aspects to this protection: domestic and global. An example of domestic protection is sulphur dioxide abatement; of course, sulphur dioxide is an international pollutant, but it has a domestic role, too. An example of global protection is carbon dioxide abatement. This is a purely global problem, and its abatement is not correlated with domestic losses. Based on empirical observations, it turns out that detecting a relationship between trade and abatement is very difficult and subject to technical econometric assumptions. If at all, these relationships are very weak, i.e. at the border of statistical significance. Yet they seem to be negative for global pollutants, such as carbon dioxide emission, and slightly positive for domestic pollutants, such as sulphur dioxide emission.

These results have intuitive interpretation. The availability of the trade option does not help traditional environmental protection. The emission of domestic pollutants – such as sulphur dioxide – is not affected by trade, at least not directly. Apparently producers who emit sulphur dioxide do not try to relocate. If they abate, it is because they are forced to do it by domestic regulations. In contrast, the availability of the trade option helps to reduce the domestic emission of global pollutants – such as carbon dioxide. But this is hardly a solution, since local abatement is linked to relocation of emission to somewhere else. It is simply yet another aspect of the carbon leakage discussed in 3.2.4.

The second issue that calls for empirical verification is the so-called Pollution Haven Hypothesis, a phenomenon linked to Summers' famous words on locating dirty industries in low income countries. This is not necessarily synonymous with the "race to the bottom" hypothesis, since – unlike the former – the latter is linked to a decision-making mechanism, rather than incomes. Statistical analyses of the Pollution Haven Hypothesis are not conclusive. There is some evidence that trade may encourage specialization in polluting goods in poor countries, but there is also some evidence to the contrary. Of course, there can be numerous examples which demonstrate that a polluting activity is found in a low income country, but also examples that high income countries specialize in pollution intensive products.

Trade is but a fragment of international exchange. Capital transfers are at least as important for environmental quality as trading in goods. It is often emphasized that all economies – and especially less developed ones – need massive capital inflows in order to meet their policy aspirations, including environmental quality. At the same time technologies embodied in foreign capital do not necessarily imply the environmental quality expected by the host country. Environmental activists have complained that so-called transition economies became a "dumping site" for corporations who transferred some of their "dirty" operations away from constituencies that are much better prepared to enforce standards (Kruszewska, 1993). While it is true that foreign direct investment added to (rather than alleviated) environmental disruption in many cases, there are no comprehensive analyses to demonstrate a capital variant of the Pollution Haven Hypothesis.

4.4.2 Policy conclusions

Trade and environment are interlinked, but specific cause-effect relationships do not allow for easy conclusions. The 1993 Swedish government report concluded that trade is neither good nor bad for the environment: competent policy makers can harness trade so as to protect the environment, while less competent ones may helplessly stand by and watch their economies drift and environments deteriorate.

In the real world, many countries are led by politicians who are not sufficiently competent . Under such circumstances, advocating for free trade is not a universally good prescription. Advocates of free trade are right when they say that trade can deliver greater wealth and consequently a better environment, but they often fail to warn that all these benign effects require conditions that may not be satisfied in a given country. Sometimes they do not feel they should educate politicians how to organize things in their countries. Moreover, it is often the case that analysts providing foreign politicians with advice on how to run their countries are accused of paternalism.

As a result – unless rooted in a more complex argument – advocating for free trade is politically controversial. A good illustration of the problem is provided by Polish attempts to limit the flow of cars damaged in accidents from its Western neighbours. Following accession to the European Union in 2004, the inflow of damaged cars (called euphemistically "used cars") increased sharply. In the peak year of 2008, the number surpassed one million. In 2012–2014 it was approximately 700,000 per year. More than half of the total are cars older than 10 years. Every few years the Polish government has tried to introduce some constraints on this flow, as it compromises road safety and imposes high waste management costs. Each time, however, the European Commission found it illegal from the point of view of internal free trade regulations. Consequently Polish regulations were withdrawn. Each time the government was advised to strengthen its enforcement capacity in order to discipline vehicle owners without discriminating against those who bought their used cars abroad. It is obvious that enforcing adequate disposal becomes more difficult when the number of potential transgressors is significantly higher, but it can be considered an "internal" problem of the government.

This is how free trade debates address environmental protection problem. They do this in a very peculiar way. Indeed, it is true that trade may help the environment. But at the same time, it can hurt the environment, unless certain conditions are met. Very often these conditions are not met, and it is obvious that they are not met. Insisting on free trade nevertheless can be defended in good faith only as a means to trigger social dissent, which may lead to establishing what has been lacking. Unconditional free trade enthusiasts can justify their claims by such an argument, but some analysts interpret this as hypocrisy.

Summary and Conclusions

Economic justification of environmental protection is based on maximizing the surplus of total benefits over total costs. Subject to standard assumptions on concavity of the former and convexity of the latter, the optimal level of environmental protection is achieved when its marginal benefit is equal to its marginal cost. Maximizing the total surplus does not imply that everybody enjoys benefits higher than costs. There can be net losers, but their joint loss is lower than the joint benefit enjoyed by net winners. Hence, the latter can compensate the former and still be better off than without environmental protection.

In a national framework there is a government who is able to enforce this optimum by means of regulations even without direct compensation. In order to achieve the optimum, the surplus has to be maximized, i.e. the winners should be able to compensate the losers, but actually compensations do not need to be paid; it is sufficient that the losers appreciate being a part of the system whose performance is optimized.

In an international framework there is no government able to enforce this optimum. As a result, in order to achieve it, compensation must be actually paid. It does not have to be pecuniary. The losers can be compensated by being offered something they appreciate, such as safety, prestige, or development opportunities.

In some agreements, where polluters and victims can be identified clearly and separated, financial compensation has been actually paid. With the Netherlands contributing to the Saline Fund, the Rhine Treaty is an excellent example of the Victim Pays Principle. The same principle has probably worked in the Colorado and Columbia cases, but compensation was not transparent. In the former, the US benefited from some parallel developments, while in the latter Canada gained additional markets for its hydroelectricity.

If there are no "spill-over" effects, i.e. if countries can fully enjoy what they paid for, adequate compensation is expected whenever all parties are rational. But "spill-over" effects are inevitable in many environmental problems. If Finland were to "buy" the Baltic Sea clean-up, say, in Latvia, then Sweden would enjoy it too, as both countries share the same sea. If Japan were to "buy" climate protection, say, by investing in lower GHG emissions in India, then Australia would enjoy it too, as both countries share the same atmosphere. Consequently no single country is willing to pay for environmental protection expecting that somebody else will do it, and the benefits will be shared. A common good – like the sea or the climate – that cannot be appropriated by any single beneficiary is called by economists a "public" good. Unwillingness to pay for such a good is called by them "free-riding". "Free-riding" is thus a curse preventing adequate protection of many goods shared by several countries.

In the case of a good which is public for a relatively small group of countries – like the Baltic Sea – triggering cooperation is easier than it is for the entire world. Nevertheless the example of the Helsinki convention demonstrates that, even for nine countries who share many common values and are relatively well developed, achieving agreement is difficult. The convention improves protection, but its ambition is below what would be justified by looking at the sea as a public good; the protection that countries undertake is largely motivated by confronting purely domestic costs with purely domestic benefits. They move towards a cleaner sea, but the pace of this process is considered disappointingly slow by many stakeholders.

In the case of a good which is public for the entire world – like the ozone layer or the global climate – the political experience is mixed. The problem of the ozone layer has been nearly solved. By contrast, the problem of climate protection is still a tremendous challenge. The key factor which makes the difference is the cost of protection. In the case of the ozone layer, costs turned out to be relatively low. Thus, "free-riding" was not considered a major issue. Potential "free-riders" would not save much by pretending that they did not care to solve the problem. Thus, acceptance of the Montreal Protocol, which established binding targets for all signatories, was universal. In the case of climate protection, the cost is very high. Thus, many countries can pretend that they do not care to solve the problem since they have more urgent needs, or that they are not guilty, or something to the same effect. As a result, the 1992 convention has not been amended with any protocol to set binding targets for countries with the highest GHG growth potential.

Of course, there are additional circumstances which made acceptance of the Montreal Protocol universal. Trade sanctions were very important. At the same time, hypothetical trade sanctions in a climate agreement are more problematic. Nonetheless, an essential reason why they were adopted in the former agreement is the cost of abatement.

Biodiversity makes an interesting case. It is a global public good, and its protection may follow either the ozone layer or the climate example. The former would imply a successful solution while the latter – a failure. The fact that biodiversity projects do not require massive costs may suggest that a more successful pattern is likely. But the reason these costs are low is that the level of ambition has been rather moderate. Should the global community contemplate a truly ambitious biodiversity protection, then costs would rise significantly, as would incentives to "free-ride". Thus, the fate of the biodiversity convention is difficult to assess.

This book concentrates on several environmental issues that have most attracted international attention, though a number of non-environmental international activities have profound environmental consequences as well. The final chapter addresses such questions.

First of all, international development assistance is examined. Its environmental component is small, but some general issues – such as "un-tying aid" – shed interesting light on environmental cooperation. Bilateralism is a logic which is applied quite frequently. Based on the Polish debt-for-environment swap, it is often argued that multilateralism (rather than bilateralism) helps environmental protection better. It promotes cost-effectiveness and goal-oriented action instead of "doing something" or "moving in the right direction", which soothes people's anxiety but does not necessarily serve the purpose. Two well-known European programmes – the Lisbon Strategy and Natura 2000 – are studied in order to analyse how large regional actions affect environmental protection both at regional and local scales.

Globalisation and international trade have become an important aspect of contemporary economic activities. Their links with environmental protection have been researched extensively. One conclusion that can be drawn from these studies is that trade is neutral with respect to the environment. More specifically, trade can be either beneficial or damaging depending on whether the trading country knows how to carry it out. This general conclusion can be questioned, however, on the grounds that, in reality, many countries do not know how to carry it out, or do not care (because of government corruption or low competence). As long as this situation does not change, claiming neutrality of trade is wrong. This does not mean that autarky is good for the environment; it is not. It simply means that environmental protection can be best served if authorities enhance their performance resulting not only in an improved understanding of people's environmental preferences, but also in improved trade arrangements.

International environmental cooperation is not only about how countries address well identified and publicised environmental problems. It is also about how to arrange cooperation in other areas so that nature is served simultaneously. Of course, explicit analyses of environmental repercussions of non-environmental actions are an important element of this. As the success of the Polish debt-for-environment swap demonstrates, building local institutions to address global environmental concerns may be more critical than short-term fully funded actions to solve a given technical problem.

International environmental cooperation can be studied from many different points of view: for instance, as a political, ethical, technical or engineering problem. The perspective adopted in this book was economic. Seen from this point of view, environmental protection is not about how to strike a balance between benefits and costs from alternative levels of ambition, but rather about how to organize cooperation between sovereign states in order to let them improve the situation. In the case of unilateral externalities, the popular *Polluter Pays Principle* may yield to the odd *Victim Pays Principle*. In the case of multilateral externalities, the challenge is how to motivate countries to overcome their natural tendency to justify decisions by confronting costs with local benefits, and to provide an appropriate public good. Avoiding "free-riding" behaviour constitutes a major challenge.

An easy assertion would be to expect avoiding such a behaviour in smaller groups of countries. Comparing the slow progress of the Baltic Sea agreements with the much faster pace of European long range transboundary air pollution demonstrates that the mechanism is more complex. Apparently the ability of leaders to inspire and convince others depends not only on their strength and relative number (*vis à vis* the entire group), but also on followers' perception of domestic benefits resulting from actions motivated by the provision of the public good.

International environmental cooperation provides a fascinating opportunity to study environmental protection as an action undertaken by sovereign agents inspired by an elite group of those who look beyond what can be perceived as their immediate benefit.

References

Ahlvik, L., P. Ekholm, K. Hyytiainen, and H. Pitkanen 2014. "An Economic-Ecological Model to Evaluate Impacts of Nutrient Abatement in the Baltic Sea", *Environmental Modelling & Software*, vol. 55, pp. 164–175.

Ahlvik, L. Y. Pavlova 2013, "A Strategic Analysis of Eutrophication Abatement in the Baltic Sea", *Environmental and Resource Economics*, vol. 56, no. 3, pp. 353–378.

Ahtiainen, H. et al. 2014, "Benefits of meeting nutrient reduction targets for the Baltic Sea – a contingent valuation study in the nine coastal states", *Journal of Environmental Economics and Policy.*

Aichele, R. and Felbermayr, G. 2012. "Kyoto and the carbon footprint of nations" *Journal of Environmental Economics and Management*, vol. 63(3), pp. 336–354.

Althammer, W. and W. Buchholz 1995, "Strategic Trade Incentives in Environmental Policy", *FinanzArchiv*, Band 52, Heft 3, pp. 293–305.

Andersson, T., C. Folke, S. Nystrom 1995, *Trading with the Environment. Ecology, Economics, Institutions and Policy*, Earthscan, London.

Ausubel, J. H. and P. E. Waggoner 2008, "Dematerialization: Variety, caution, and persistence", PNAS, vol. 105, no. 35, pp. 12774–12779 (available at www.pnas.org/cgi/doi/10.1073/pnas.0806099105).

Babiker M. 2005, "Climate change policy, market structure, and carbon leakage", *Journal of International Economics*, Vol. 65, pp. 421–445.

Bajsarowicz J., and T. Zylicz 1993, "Environmental Reform: The Polish Model", *The Wall Street Journal Europe*, July 1.

Barde, J. P. 1992, *Economie et Politique de l'Environnement*, Presses Universitaires de France, Paris [Environmental economics and policy].

Barnett, H. J., Ch. Morse 1963, *Scarcity and Growth. The Economics of Natural Resource Availability*, Johns Hopkins, Baltimore.

Barrett, S. 1994, "Self-Enforcing International Environmental Agreements", *Oxford Economic Papers*, vol. 46, pp. 878–894.

Baumol, W. J., W. E. Oates 1988, *The Theory of Environmental Policy*, Cambridge University Press, Cambridge.

Bhagwati, J., T. N. Srinivasan 1996, "Trade and the Environment: Does Environmental Diversity Detract from the Case of Free Trade?", in: *Fair Trade and Harmonization: Prerequisites for Free Trade?*, J. Bhagwati and R. E. Hudec (eds.), Cambridge, MIT Press, p. 159–223.

Boehringer C., C. Fischer, K. E. Rosendahl 2010, "The Global Effects of Subglobal Climate Policies", *RFF Discussions Paper* 48, Resources for the Future.

Braend, T. 2014, "CCS no more in Norway", *Acid News*, No. 1, March.

Burniaux J.-M., J. Chateau, R. Dellink, R. Duval, S. Jamet 2009, "The Economics of Climate Change Mitigation: How to Build the Necessary Global Action in a Cost-Effective Manner", *OECD Economics Department Working Papers* 701.

Burniaux, J.-M. and J. Oliveira Martins 2012, "Carbon leakages: a general equilibrium view", *Economic Theory*, Vol. 49, Iss. 2, pp. 473–495.

Carbone, J., C. Helm, T. Rutherford 2009, "The case for international emission trade in the absence of cooperative climate policy," *Journal of Environmental Economics and Management* 58(3), pp. 266–280.

Carson, R. T., and Czajkowski, M. 2014. "The Discrete Choice Experiment Approach to Environmental Contingent Valuation", in: *Handbook of choice modelling*, S. Hess and A. Daly, eds., Edward Elgar, Northampton, MA.

CBD (Convention on Biological Diversity) 2010, *Decisions of COP-10* (accessed at http://www.cbd.int/decision/cop/default.shtml?id=12313).

Chambers, P. E., R. Jensen and J. C. Whitehead 1996, "Debt-for-nature swaps as noncooperative outcomes", *Ecological Economics*, vol. 19, pp. 135–146.

Chander, P., and H. Tulkens 1992, "Theoretical foundations of negotiations and cost sharing in transfrontier pollution problems, *European Economic Review* vol. 36, pp. 388–398.

Coase, R. H. 1960, "The Problem of Social Cost", *The Journal of Law and Economics*, 3(1), pp. 1–44.

Cody, B. 1988, *Debt-for-Nature Swaps in Developing Countries*, Congressional Research Service Report 88–647 ENR, Washington, D.C.

Committee on Climate Change 2013, *Reducing the UK's carbon footprint*, London.

Communication from the European Commission to the European Parliament, the Council, the European Economic and Social Committee and the Committee of the Regions 2010, *European Union Strategy for Danube Region*, COM(2010) 715 final, Brussels.

Da Rosa, J. E. 1983, "Economics, politics, and hydroelectric power: the Parana River Basin", *Latin American Research Review*, vol. VXVIII(3), pp. 77–107.

Dasgupta, S., B. Laplante, H. Wang, and D. Wheeler 2002, "Confronting the Environmental Kuznets Curve", *The Journal of Economic Perspectives*, vol. 16(1), pp. 147–168.

Diamantoudi, E. & E. S. Sartzetakis 2006, "Stable International Environmental Agreements: An Analytical Approach", *Journal of Public Economic Theory*, vol. 8(2), pp. 247–263.

Dudek, D., A. Golub, E. Strukova 2001, "Ancillary Benefits of Reducing Greenhouse Gas Reduction in Transitional Economies", Environmental Defense Fund, New York.

EBRD (European Bank for Reconstruction and Development) 2011, *The low carbon transition. Special Report on Climate Change*, London.

EC (European Commission) 2011, *Communication from the Commission to the European Parliament, the Council, the European Economic and Social Committee and the Committee of the Regions. Roadmap to a Resource Efficient Europe*, Brussels, COM(2011) 571 final

EcoFund 1992a, *Operating Procedures Manual*, Warsaw.

EcoFund 1992b, *Project Procedures Manual*, Warsaw.

EEA (European Environment Agency) 2013, *Trends and Projections in Europe 2013. Tracking Progress towards Europe's climate and energy targets until 2020*, Copenhagen.

EEA (European Environment Agency) 2014, "Sulphur dioxide SO2 emissions (APE 001)", Copenhagen. Available *on-line* at http://www.eea.europa.eu/data-and-maps/indicators/eea-32-sulphur-dioxide-so2-emissions-1/assessment-3.

Ekins, P., C. Folke, R. Costanza 1994, "Trade, Environment and Development: The Issues in Perspective", *Ecological Economics*, vol. 9(1).

Els, F. 2014, China's rare earth industry will look very different in 5 months, Mining.com http://www.mining.com/chinas-rare-earth-industry-will-look-very-different-at-the-end-of-the-year-53315/.

Environment Canada, 1988, *Canadian achievements in controlling acidic emissions*, Ottawa

Facing the Challenge. The Lisbon strategy for growth and employment 2004, Report from the High Level Group chaired by Wim Kok, European Commission, Brussels.

Fisher, A. C., and J. V. Krutilla 1985, "Economics of Nature Preservation", in: A. V. Kneese, and J. L. Sweeney (eds.), *Handbook of Natural Resource and Energy Economics*, North Holland, Amsterdam, vol. I, pp. 165–189.

Folmer, H., P. van Mouche, and S. Ragland 1993, "Interconnected Games and International Environmental Problems", *Environmental and Resource Economics*, vol. 3(4), pp. 313–335.

Frankel, J. 2008, "Environmental Effects of International Trade", *Expert Report No. 31 to Sweden's Globalisation Council*, Stockholm.

Frijters, I. D. and J. Leentvaar 2003, *Rhine Case Study*. Technical documents in hydrology, no. 17. Paris: UNESCO International Hydrological Programme, (Rep. No. SC/2003/WS/54).

GATT (General Agreement on Tariffs and Trade) 1947, Text of the treaty, available at http://www.wto.org/english/docs_e/legal_e/gatt47_01_e.htm.

Gerlagh, R. and O. Kuik 2007, "Carbon Leakage with International Technology Spillovers", *FEEM Working Papers* 48, Fondazione Eni Enrico Mattei.

Grazia, C., F. Gusdorf, A. Hammoudi 2014, "Climate Change, Heterogeneities, and Stability of International Fiscal Harmonization", *Environmental and Resource Economics*, Vol. 58, No. 4, pp. 579–603.

Gren, I.-M., C. Folke, S. Langaas, M. Sandström, T. Söderqvist, F. Wulff, F. 1995, *Costs and Benefits from Nutrient Reductions to the Baltic Sea*. The Beijer Institute, The Royal Swedish Academy of Sciences, Stockholm, Mimeo.

Griffith-Jones, S. 1992, "Conversion of Official Bilateral Debt: The Opportunities and the Issues", paper prepared for the World Bank Development Economics Conference, Washington, D.C., April 30-May 1.

Grubb, M. 1990, "The Greenhouse Effect: Negotiating Targets", *International Affairs*, 66(1), January, pp. 67–89.

GUS (Central Statistical Office) 1997, *Ochrona Srodowiska 1997*, Warsaw [Environmental Yearbook 1997].

Haas, P. M. 1992, "Banning Chlorofluorocarbons: Epistemic Community Efforts to Protect Stratospheric Ozone", *International Organization* 46(1), Winter, pp. 187–224.

Hanley, N., J. Shogren, and B. White, 2007, *Environmental Economics in Theory and Practice*, Palgrave, London.

Hansen, S. 1989, "Debt for nature swaps – Overview and discussion of key issues", *Ecological Economics*, vol. 1, pp. 77–93.

Hasler, B., J. C. R. Smart, A. Fonnesbech-Wulff, H. E. Andersen, H. Thodsen, G. Mathiesen, E. Smedberg, C. Göke, M. Czajkowski, A. Was, K. Elofsson, C. Humborg, A. Wolfsberg, F. Wulff 2014, "Hydro-economic modelling of cost-effective transboundary water quality management in the Baltic Sea", *Water Resources and Economics*. Available on-line at: http://dx.doi.org/10.1016/j.wre.2014.05.001.

Helcom 2010, *Ecosystem Health of the Baltic Sea*, Baltic Sea Environment Proceedings No. 122, Helsinki.

Helcom 2011, *Fifth Baltic Sea Pollution Load Compilation (PLC-5)*, Baltic Sea Environment Proceedings No. 128, Helsinki.

Highton, N. H. and M. J. Chadwick 1982, "The Effects of Changing Patterns of Energy Use on Sulfur Emissions and Deposition in Europe", *Ambio* 6, pp. 325–330.

Hu, Tao 2004, "Policy Implications of Co-benefit in China: Linking Air Pollution and GHGs", Policy Research Center for Environment & Economy SEPA (accessed through internet on February 7, 2008 at http://www.ap-net.org/docs/14th_seminar/hutao.pdf).

Hufbauer, G. C., S. Charnovitz, J. Kim 2009, *Global Warming and the World Trading System*, Peterson Institute for International Economics, Washington, DC.

Hundley, N. 2009, *Water and the West: The Colorado River Compact and the Politics of Water in the American West*, University of California Press.

IPCC (Intergovernmental Panel on Climate Change) 2007a, *Contribution of Working Group I to the Fourth Assessment Report of the Intergovernmental Panel on Climate Change*, Solomon, S., D. Qin, M. Manning, Z. Chen, M. Marquis, K.B. Averyt, M. Tignor and H.L. Miller (eds.), Cambridge University Press.

IPCC (Intergovernmental Panel on Climate Change) 2007b, *Contribution of Working Group II to the Fourth Assessment Report of the Intergovernmental Panel on Climate Change*, M.L. Parry, O.F. Canziani, J.P. Palutikof, P.J. van der Linden and C.E. Hanson (eds.), Cambridge University Press.

IPCC (Intergovernmental Panel on Climate Change) 2007c, *Contribution of Working Group III to the Fourth Assessment Report of the Intergovernmental Panel on Climate Change*, B. Metz, O.R. Davidson, P.R. Bosch, R. Dave, L.A. Meyer (eds.), Cambridge University Press.

IPCC (Intergovernmental Panel on Climate Change) 2009, *IPCC Programme and Budget* (Submitted by the Secretariat), Geneva.

IPCC (Intergovernmental Panel on Climate Change), 2010, *Guidance Note for Lead Authors of the IPCC Fifth Assessment Report on Consistent Treatment of Uncertainties*, Bonn.

IPCC (Intergovernmental Panel on Climate Change) 2013, *Climate Change 2013. The Physical Science Basis. Working Group I Contribution to the Fifth Assessment Report of the IPCC*, Cambridge University Press, Cambridge and New York.

IPCC (Intergovernmental Panel on Climate Change) 2014a, *Climate Change 2014. Impacts, Adaptation, and Vulnerability. Working Group II Contribution to the Fifth Assessment Report of the IPCC*, Cambridge University Press, Cambridge and New York.

IPCC (Intergovernmental Panel on Climate Change) 2014b, *Climate Change 2014: Mitigation of Climate Change. Contribution of Working Group III to the Fifth Assessment Report of the IPCC*, Cambridge University Press, Cambridge and New York.

Jevons, W. S. 1865, *The Coal Question*, Macmillan.

Jobert, T., F. Karanfil, A. Tykhonenko 2010. "Convergence of per capita carbon dioxide emissions in the EU: Legend or reality?", *Energy Economics* vol. 32, pp. 1364–1373.

Johansson, P.-O., B. Kristrom 2007, "On a clear day you might see an environmental Kuznets curve", *Environmental and Resource Economics*, vol. 37(1), pp. 77–90.

Joint Statement on Guiding Principles for the Development of Inland Navigation and Environmental Protection in the Danube River Basin 2007, Vienna.

Kiuila, O. 2000, „Skutki realizacji Drugiego Protokolu Siarkowego w Polsce", in: *Ekonomia i Srodowisko* 17(2), pp. 87–99 [Effects of the Second Sulphur Protocol in Poland].

Kiuila, O., K. Wojtowicz, T. Zylicz, L. Kasek 2014, "Economic and environmental effects of unilateral climate actions". *Mitig Adapt Strateg Glob Change*. DOI 10.1007/s11027-014-9597-9.

KPMG 1997, *Review of the Polish EcoFund*, KPMG, Warsaw.

Kruszewska, I. 1993, "Open Borders, Broken Promises. Privatization and Foreign Investment: Protecting the Environment through Contractual Clauses", Greenpeace, Amsterdam.

Kukla-Gryz, A. 2009, "Economic Growth, International Trade and Air Pollution: A Decomposition Analysis", *Ecological Economics*, vol. 68(5), pp. 1329–1339.

Lang, John Temple. 1993, Biological Conservation and Biological Diversity, in: Sjöstedt, 1993, pp. 171–188.

Lashof D. & D. Tirpak 1989, *Policy options for stabilizing global climate*, US EPA, Washington, DC.

Lesser, W. H. and A. Krattiger 2007, "Valuation of Bioprospecting Samples: Approaches, Calculations, and Implications for Policy-Makers", in: *Intellectual Property Management in Health and Agricultural Innovation: A Handbook of Best Practices* (eds. A. Krattiger, R. T. Mahoney, L. Nelsen, *et al.*). MIHR, Oxford, and PIPRA, Davis. Available online at www.ipHandbook.org.

Lovejoy, T. J. 1984, "Aid Debtor Nations' Ecology", *The New York Times*, October 4, p. A 31.

Lundberg, F., G. Boye Olesen 2014, "150 ways to cut GHG emissions", *Acid News*, No. 2, June, pp. 1 and 4–5.

Magnusson, K. 2014, "Microlitter and other microscopic anthropogenic particles in the sea area off Rauma and Turku, Finland", IVL Svenska Miljöinstitutet, Gothenburg.

Mäler, K.-G. 1989, "The acid rain game", in: H. Folmer and E. van Ierland, eds., *Valuation Methods and Policy Making in Environmental Economics*, Elsevier, Amsterdam, pp. 221–252.

Mäler, K-G. 1990, "International Environmental Problems", *Oxford Review of Economic Policy*, vol. 6, no. 1, pp. 80–108.

Mahony, R. 1992, "Debt-for-Nature Swaps. Who Really Benefits?", *The Ecologist*, Vol. 22, No. 3, May/June, pp. 97–103.

Malthus, T. R. 1789, *An essay on the principle of population, as it affects the future improvement of society, with remark on the speculations of Mr. Godwin, M. Condorcet and other writers*, J. Johnson, in St. Paul's Church-yard, London http://www.econlib.org/library/Malthus/malPop.html.

Markowska, A. and T. Zylicz 1999, "Costing an international public good: The case of the Baltic Sea", *Ecological Economics* 1999, vol. 30, pp. 301–316.

Marshak, S. 2007, *Essentials of Geology*, W. W. Norton.

McCaffrey, S. C. 2008, *Convention on the Law of the Non-Navigational Uses of International Watercourses*, United Nations Audiovisual Library of International Law, New York. Available at www.un.org/law/avl.

Meadows, D. H. 1988, "The Limits to Growth Revisited", in: P. R. Ehrlich and J. Holdren (eds), *The Cassandra Conference: Resources and the Human Predicament*, Texas A&M University Press, College Station, pp. 257–270.

Mickwitz, P. 1998, *Implementation of Key Environmental Principles. Experiences from the Protection of the Baltic Sea*, Nordic Council of Ministers, Copenhagen.

Mikdashi, Z. 1974, "Collusion Could Work", *Foreign Policy*, 14, Spring, pp. 57–68.

Mill, J. S. 1848, *Principles of Political Economy*, John W. Parker, London http://www.econlib.org/library/Mill/mlP.html.

Mylona, S. 1996, "Sulphur dioxide emissions in Europe 1880–1991", *Tellus*, 48B, 662–689.

Mylona, S. 1997 "Corrigendum to Sulphur dioxide emissions in Europe 1880–1991", *Tellus*, 49B, 447–448.

Nagoya Protocol on Access to Genetic Resources and the Fair and Equitable Sharing of Benefits Arising from Their Utilization to the Convention on Biological Diversity 2010, Montreal.

NAS (National Academy of Science) 1992, *Policy Implications of Greenhouse Warming. Mitigation, Adaptation, and the Science Base*, Washington, DC.

NASA (National Atmospheric and Space Authority) 2012, *Headquarters release* No. 12–020.

Nordhaus, W. 2007, *The Challenge of Global Warming: Economic Models and Environmental Policy*, Yale University, New Haven.

Nowicki, M. 1991, "Redirecting Debt Service for Environmental Recovery Purposes". Memorandum, Warsaw, Ministry of Environment.

OECD (Organization for Economic Cooperation and Development) 1981, *Transfrontier Pollution and the Role of States*, Paris.

OECD (Organization for Economic Cooperation and Development) 1998, *Swapping Debt for the Environment. The Polish EcoFund*, OECD, Paris.

OECD (Organization for Economic Cooperation and Development) 2008, *Development Aid at a Glance 2008. Statistics by Region*, Paris.

OECD (Organization for Economic Cooperation and Development) 2014, Statistical tables available on-line at http://www.oecd.org/statistics/.

Ohlin, B. 1967, *Interregional and International Trade*, Cambridge, Harvard University Press.

Oloo, A. 2007, "The Quest for Cooperation in the Nile Water Conflicts: A Case for Eritrea", *African Sociological Review* Vol. 11(1), pp. 95–105.

Olson, M., Jr. 1963, "American Materials Policy and the 'Physiocratic Fallacy'", *Orbis* VI(4), Winter, pp. 670–688.

Paas, T., F. Schlitte 2007, "Regional Income Inequality and Convergence Processes in the EU-25", *HWWI Research Papers* 1–11, Hamburg Institute of International Economics.

Porter, M., and C. van der Linde 1995, "Toward a New Conception of the Environment-Competitiveness Relationship", *Journal of Economic Perspectives*, vol. 9(4), pp. 97–118.

Rauscher, M. 2005, "International Trade, Foreign Investment, and the Environment", chapter 27 in: *Economywide and International Environmental Issues*, Volume 3 of *Handbook of Environmental Economics* (K.-G. Mäler and J. R. Vincent, eds.), pp. 1403–1456.

Republic of Poland 1991a, *Government Initiative on Environmental Fund*, Warsaw.

Republic of Poland 1991b, *Government Initiative on Environmental Fund. Report on the Conference on Polish Debt-for-Environment Programme*, Oslo-Warsaw.

Republic of Poland 1992a, *Polish Environment Foundation. EcoFund Charter*, Warsaw.

Republic of Poland 1992b, *Government Initiative on Environment Fund. EcoFund Concept Paper*, Warsaw.

Republic of Poland 1992c, *Government Initiative on Environment Fund. Role and Potential Impact of the EcoFund*, Warsaw.

Ribbe, L. 1994, "Der Ecofund in Polen", *Euronatur Hintergrund* [The EcoFund in Poland].

Ricardo, D. 1817, *On the Principles of Political Economy and Taxation*, John Murray, Albemarle-Street, London http://www.econlib.org/library/Ricardo/ricP-Cover.html.

Ritter, H., M. Schopf 2014, "Unilateral Climate Policy: Harmful or Even Disastrous?", *Environmental and Resource Economics*, vol. 58, pp. 155–178.

RIVM (Dutch National Institute for Public Health and the Environment) 1999, "European Environmental Priorities. Report to the European Commission with Annexes", Brussels.

Rohwedder, C. 1993, "Poland Sets Pace for Cleanup in the East, And 'Debt-for-Nature' Swap Adds Leverage", *The Wall Street Journal Europe*, January 22.

Sebenius, J. K. 1993, "The Law of the Sea Conference: Lessons for Negotiations to Control Global Warming", in: Gunnar Sjöstedt (1993), pp. 189–216.

Shabecoff, P. 1987, "Bolivia: To Protect Lands in Swap for Lower Debt", *The New York Times*, July 14.

Shaw, R. W. 1993, "Acid Rain Negotiations in North America and Europe: A Study in Contrast", in: Gunnar Sjöstedt *et al.* (1993), pp. 84–109.

Simpson, R. D., M. A. Toman, and R. U. Ayres (eds.) 2005, *Scarcity and Growth Revisited*, RFF Press, Washington.

Sinn, H.-W. 2008, "Public policies against global warming: a supply side approach", *International Tax and Public Finance*, Vol. 15, pp. 360–394.

Sjöstedt, G. (ed.) 1993, *International Environmental Negotiation*, Sage, London.

Sjöstedt, G., U. Svedin 1993, "An Introduction to International Environmental Negotiations", in: Gunnar Sjöstedt *et al.* (1993), pp. 7–11.

Sjöstedt, G., U. Svedin, B. Hägerhäll Aniansson (eds.) 1993, *International Environmental Negotiations. Process, Issues and Contexts*, Swedish Council for Planning and Coordination and Research, Stockholm.

Smith, A. 1776, *An Inquiry into the Nature and Causes of the Wealth of Nations*, London, Strahan and Cadell http://www.econlib.org/library/Smith/smWN.html.

Smith, S. J., J. van Aardenne, Z. Klimont, R. Andres, A. Volke, and S. Delgado Arias 2010, "Anthropogenic sulfur dioxide emissions: 1850–2005", *Atmos. Chem. Phys. Discuss.*, vol.10, pp. 16111–16151, doi:10.5194/acpd-10–16111–2010.

Sneddon, C., and C. Fox 2007 "Transboundary river basin agreements in the Mekong and Zambezi basins: enhancing environmental security or securitizing the environment?", *International Environmental Agreements*, vol. 7, pp. 237–261.

Stanley, M. 2012, *Harnessing The Power: Voices from Two Rivers of the Peace and Columbia*. Vancouver: Douglas & McIntyre Publishers Inc.

Starr, J. R. 1991, "Water Wars", *Foreign Policy*, 82, Spring, pp. 17–36.

Stavins, R. 2014, "Is the IPCC Government Approval Process Broken?", in: *An Economic View of the Environment*, Harvard University. Available at http://

www.robertstavinsblog.org/2014/04/25/is-the-ipcc-government-approval-process-broken-2/.

Stiglitz, J. 1974, "Growth with Exhaustible Natural Resources: Efficient and Optimal Growth Paths", *Review of Economic Studies*, Vol. 41, Iss. 5, pp. 123–137.

Széll, P. 1993, "Negotiations on the Ozone Layer", in: Sjöstedt (1993), pp. 31–47.

Tol, R. 2005, "The marginal damage costs of carbon dioxide emissions: an assessment of the uncertainties", *Energy Policy*, vol. 33, pp. 2064–2074.

Tol, R. S. J. 2014, *Climate Economics. Economic Analysis of Climate, Climate Change and Climate Policy*, Edward Elgar, Cheltenham.

UNCTAD (United Nations Conference on Trade and Development) 1992, *Conversion of Official Bilateral Debt*, GE.92–55494, Geneva.

UN ECE (United Nations Economic Commission for Europe) 1994, *Recommendations to ECE governments on the financing of environmental policies*, Geneva ECE/CEP/4 ECE/EC.AD/45.

Vahtera, E., D. J. Conley, B. G. Gustafsson, H. Kuosa, H. Pitkänen, O. P. Savchuk, T. Tamminen, M. Viitasalo, M. Voss, N. Wasmund, F. Wulff 2007, "Internal ecosystem feedbacks enhance nitrogen-fixing cyanobacteria blooms and complicate management in the Baltic Sea", *Ambio* Vol. 36, Iss. 2, pp. 186–194.

Varhelyi, G. 1985, "Continental and Global Sulfur Budgets – I. Anthropogenic SO2 emissions", *Atmospheric Environment* 7/85, pp. 1029–1040.

Vestreng, V., G. Myhre, H. Fagerli, S. Reis, and L. Tarrason 2007, "Twenty-five years of continuous sulphur dioxide emission reduction in Europe", *Atmos. Chem. Phys.*, vol. 7, pp. 3663–3681 Available at www.atmos-chem-phys.net/7/3663/2007/.

WTO, 2014, China – measures related to the exportation of rare earths, tungsten, and molybdenum. AB-2014–3, AB-2014–5, AB-2014–6. Reports of the Appellate Body https://docs.wto.org/dol2fe/Pages/FE_Search/FE_S_S006.aspx?Query=%28@Symbol=%20wt/ds431/ab/r*%20not%20rw*%29&Language=ENGLISH&Context=FomerScriptedSearch&languageUIChanged=true#.

Wulff, F. et al. 2014, "Reduction of Baltic Sea Nutrient Inputs and Allocation of Abatement Costs Within the Baltic Sea Catchment", *Ambio* vol. 43, pp. 11–25.

Zacher M. W. and J. G. McConnell, 1990, "Down to the Sea with Stakes: The Evolving Law of the Sea and the Future of Deep Seabed Regime", *Ocean Development and International Law*, 11 (3/4), pp. 265–295.

Zylicz, T. 1985, „Emisja i transfer zanieczyszczen siarkowych w Europie. Ekologiczne i ekonomiczne przeslanki rozwiazania problemu", *Ekonomista* 1–2/1985, pp. 157–184 [Emissions and transfers of sulphur dioxide in Europe. Ecological and economic aspects of solving the problem].

Zylicz, T. 1992, *Debt-for-Environment Swaps: The Institutional Dimension*, Beijer Discussion Paper No. 18 The Royal Swedish Academy of Sciences, Stockholm.

Zylicz, T. 2000, *Costing Nature in a Transition Economy. Case Studies in Poland*, Edward Elgar, Chelthenham.

Zylicz, T. and M. Czajkowski 2009, "Sustainability and co-benefits of climate protection", in: K. D. John, and D. T. G. Ruebbelke (eds.), *Benefits of Environmental Policy*, Routledge, London, pp. 24–35.

Polish Studies in Economics

Edited by Ryszard Kokoszczyński

Volume 1 Wojciech Grabowski: Employee Stock Options, Payout Policy, and Stock Returns. Shareholders and Option Holders in Large U.S. Technology Corporations. 2012.

Volume 2 Jan Jakub Michałek / Michał Brzozowski / Andrzej Cieślik (eds.): The Relationship between Human Development, Exports and Foreign Direct Investments in Emerging Europe. 2013.

Volume 3 Tomasz Żylicz: The Economics of International Environmental Cooperation. 2015.

www.peterlang.com

www.ingramcontent.com/pod-product-compliance
Ingram Content Group UK Ltd.
Pitfield, Milton Keynes, MK11 3LW, UK
UKHW021842100926
8556IPUK00019B/993

9 783631 652336